Social Dimensions of Contemporary Environmental Issues:
International Perspectives

Social Dimensions of Contemporary Environmental Issues: International Perspectives

edited by

Peter Ester
and
Wolfgang Schluchter

Tilburg University Press 1996

Cover design: Martin van Kemenade

© Tilburg University Press 1996

ISBN 90-361-9507-1

No part of this publication may be reproduced or transmitted in any form or by any means, electronic or mechanical, including photocopy, recording, or any information storage and retrieval system, without permission from the copyright owner.

Contents

Preface ix

Part I Science, Technology and the Environment

1. The Trap of Science and Technology 1
Pieter Tijmes

1. Introduction 1
2. Rival interpretation 2
3. Ambiguity of science 3
4. Science without norms 5
5. The reconstructive power of science 5
6. Economics and the world of daily life 7
7. Environment corrections 8
8. In retrospect 11
References 13
Notes 14

2. Capitalism, Nature and Modernity. A Realist Perspective 15
Pieter Dickens

1. Introduction 15
2. Contemporary eco-Marxism: capitalism as the wrecker of nature? 16
3. The biology of organisms and organism-environment relations: recovering a lost tradition 22
4. Human society and nature: the division of labour as a key obstacle 27
5. Conclusions 29
References 31
Notes 33

3. Fragmentariness and Interdisciplinarity in Linking Ecology and the Social Sciences 35
Andrej Kirn

1. Introduction 35
2. Environmental problems and theoretical/paradigmatic challenges 38
3. Just linkage, or theoretical integration and unification as well? 44
4. Evolution and technological inevitability of environmental problems 46
5. Interdisciplinarity and multidisciplinarity for sustainable development 48

6. Conclusions	49
References	51
Notes	53

4. The Establishment of an International Regime: the Case of Stratospheric Ozone Depletion   55
August Gijswijt

1. Introduction	55
2. Rapidly increasing problem awareness; expansion and differentiation of knowledge (1974-1980)	56
3. Fading concern; empirical research as a selection mechanism for the accuracy of rival theories and hypotheses (1981-1986)	58
4. The realization of the ozone regime (1986-1987)	60
5. Tightening up of the ozone regime (1987-1992)	64
6. The background to social reactions to the ozone question	66
References	75
Notes	78

Part II Environmentalism

5. Globalization, Environmental Awareness, and Ecological Behavior Shown at the Example of the Federal Republic of Germany   81
Wolfgang Schluchter and Andreas Metzner

1. Introduction	81
2. The evolution of environmental awareness as a learning process	83
3. The relationship between environmental awareness and behavior	86
4. Conclusions	94
References	96

6. The Evolution of the Soviet/Russian Ecological Movement: Political Trends   99
Oleg Yanitsky and Irene Khalyi

1. Introduction	99
2. The emergence of the ecological movement	99
3. Perestroika: 'ecological solidarity' and mass protests of 1988-1991	102
4. Elections to the all-union and republican parliaments, 1989-1990	104
5. 1990-1992: cooperation with the authorities; the beginning of the movement's deep transformations	105
6. 1993-1995: new obstacles and missed opportunities	106

| 7. | Conclusions | 109 |
| References | | 111 |

7. Support for Environmentalism During a Major Recession — 113
Elim Papadakis

1.	Introduction	113
2.	The 1993 election campaign	115
3.	The environment as a political issue	118
4.	The environment versus development	120
5.	Sustainable development	128
References		130
Notes		131

Part III   Public Opinion and the Environment

8. Global Environmental Concern: A Challenge to the Post-materialism Thesis — 133
Riley Dunlap and Angela Mertig

1.	Introduction	133
2.	Methodology	134
3.	Results	140
4.	Conclusions	155
References		158
Notes		160

9. Individual Change and Stability in Environmental Concern: The Netherlands 1985-1990 — 165
Peter Ester, Loek Halman and Brigitte Seuren

1.	Introduction	165
2.	Hypotheses	169
3.	Data and measurement instruments	170
4.	Individual changes in environmental concern	174
5.	Environmental switchers and stable types	178
6.	Characteristics of environmental types	180
7.	Conclusions	183
References		185
Notes		187

10. Farmers' Attitudes to Environmental Issues: An Australian Study 189
Alan Black and Ian Reeve

1.	Introduction	189
2.	Theoretical considerations	190
3.	Survey design and sample	195
4.	Results of analysis	197
5.	Conclusions	207
References		209
Notes		213

11. Methodological Problems of Measurement of Ecological Attitudes and Comparison of Survey Data 215
Vladimir Rukavishnikov

1.	Introduction	215
2.	Advantages and limitations of surveys' information about ecological attitudes	216
3.	Four basic research questions for comparitive studies	219
4.	Conclusions	225
References		227

About the authors 229

Preface

During the congress of the International Sociological Association, held in Bielefeld, Germany in June 1994, the Working Group on Environment and Society put forward a proposal to make a number of the congress contributions available to a wider public in order to stimulate debate on the issues discussed therein.
Technologically advanced societies now find themselves in a crisis of transition, one of whose main characteristics is that strategies based on the model of a quantitative increase in the living standard, which up to present have been pursued with success, can no longer be implemented under conditions of increasing environmental depletion. Central to the Industrial Revolution of the last century was a technomorphous development which attempted to tackle structural problems with the very means that were responsible for negative environmental effects and thus reduction in environmental quality in the first place.
Crises of transition may also be discerned in less technologically developed societies; however, due to the almost infinite availability of human resources they are not so apparent. Nevertheless, it should be noted that when such societies do embark on a course of Western industrialization the same problems arise, often in acuter and more exacerbated forms. The globalization of environmental problems and their transnational dimensions have created a state of awareness in Western societies that has moved the focus of debate away from the apportioning of "blame" about causes and effects of environmental degradation to concentrate more on the intermeshing of victims and culprits.
New hazards are arising, both in individual countries and societies and across borders. These hazards increase in amplitude as natural resources for human usage decrease in quantity and quality. In many Western countries there is now concentrated debate on how best to abrogate and correct such erroneous courses of development whose consequences will indeed be horrendous.
In countries of the Third World the debate is centered on finding solutions to social problems connected with population escalation. A constant factor in this debate is the uncritical attitude displayed with regard to the inevitability of Western-style technological development. This attitude is fostered by the technologically advanced countries whether through the "force majeure" of their economic globalization strategies or through the unspoken force of their own example held up high to people in the Third World.
The globalized nature of environmental problems enables individual countries to "pass the buck" for environmental degradation easily onto other parties. It further poses a serious impediment to any single country that wishes to launch effective measures against increasing environmental degradation on its own.
No country can act effectively on the local level as long as there is a lack of processes of consensus which seek to reconcile the demands of disparate development paradigms and which are grounded in a worldwide awareness of

the dangerous potential of the human race for complete self-destruction. "Thinking globally and acting locally" is the challenge that people will have to take up in the future.

It is imperative to initiate a process for consensus on a world scale. The World Summit on the Environment held in Rio de Janeiro in 1992 is a preliminary expression of such a process; however unequal rates of economic and social development continue to pose serious obstacles in the way of the search for consensus, and as yet we may look in vain for signs of effective action on a global scale - perhaps, indeed, they will never come about. Nationalistic, short-term and selfish interests unfortunately still outweigh the global, long-term collective interests of humanity. The close interconnections of world market economic relations and the increasingly technological nature of societies in a deep crisis of transformation can only serve to exacerbate environmental problems if a transnational consensus is not achieved. The foundation of such a consensus must lie with ideas of equality and human rights which include the right to an intact, sustainable environment. International efforts to solve global environmental problems will have to produce fundamental changes in Western oriented life styles and are bound to bring about (deep) social dislocation(s).

The social sciences must monitor and accompany these processes more than they have done in the past, their findings can offer a contribution to a much needed change of paradigm. It is precisely such a goal that the present reader intends to serve.

The content of this book is structured along three parts. *Part I* offers a general reflection on the interrelationships between science, technology and the environment. In the first chapter *Pieter Tijmes* shows how science and technology - strongly reinforced by economics - have shaped the modern worldview. But this alliance of the homo faber and the homo oeconomicus is not an unproblematic one. His discussion of this alliance ends with a criticism of technocracy and with a plea for ascesis. In the second chapter *Peter Dickens* analyses the way in which a highly complex division of labor in modern society has alienated human beings from nature. More precisely this form of alienation is seen as a failure of people to understand their relations with nature. He proposes a new alliance between the social and life sciences, one which develops an almost lost approach to understanding the relations between organisms and the environment. In the next chapter *Kirn* argues that ecology as a natural science is less and less able to observe processes in the biosphere independently of human activities. Input from social scientific theory and research is much needed for a full comprehension of environmental problems. He shows that the theoretical formulation and practical realization of the concept of an ecologically sustainable society demands interdisciplinarity in science, particularly as regards the social, natural and technical sciences. In the fourth chapter *Gijswijt* reconstructs international policy cycles and negotiation processes with respect to the ozone question. He analyzes the process of problem definition and problem awareness and the establishment of what is called "the ozone regime". His chapter focusses on how international ozone policy was and is affected by environmentalists, industry and public opinion.

Part II of this book contains contributions to the study of environmentalism in various countries. In the chapter by *Schluchter and Metzner* the rise of environmental movements in the Federal Republic of Germany is analyzed. They observe the unfolding of a new ecological paradigm which stresses restricted use of scarce resources and the improvement of environmental quality. At the same time, however, it is shown that the political system is lagging behind as it limits itself to merely symbolic environmental policies based on non-committal declarations and intentions. In chapter six *Yanitsky and Khalyi* describe the failure of the environmental movement in Russia to become a crucial and significant social movement. This failure is explained in terms that the environmental movement has not adequately used existing political opportunities, was mainly oriented towards correcting 'mistakes' by the governmental regime, and was less directed to political confrontations than other new social movements. In chapter seven *Papadakis* reports on the role played by environmental issues during a major recent recession in Australia. Though these issues played a less prominent role during election campaigns, the environment paradoxically remains an issue of great concern among the public at large. He explains this paradox by hypothesizing that a new perception of the relationship between environment and development has been adopted that sees environmental protection as complimentary rather than fundamentally opposed to economic development.

Part III reports on recent studies on public opinion and the environment. In chapter eight *Dunlap and Mertig* present findings of a major international study on public concern for environmental quality in 24 nations. Their findings indicate that environmental concern is widely shared in both wealthy and poor nations. Contradicting conventional wisdom as well as the post-materialism thesis, overall national affluence is more often negatively rather than positively related to citizen concern for environmental quality. In the next chapter *Ester, Halman and Seuren* show that in the Netherlands the overall stable trend in environmental concern seems to mask attitudinal turbulency at the individual level. Using panel data they find that approximately half of the Dutch citizens switched their position with respect to environmental issues. Furthermore they find that a majority of the Dutch population appears to be on the midway between a rejection of the old industrial worldview and the adoption of a new post-industrial wordlview. In their chapter on farmers' attitudes to rural environmental issues in Australia *Black and Reeve* show how these attitudes vary across different agricultural industries and relate to social characteristics of farmers, as well as to characteristics of their farms. One of their conclusions is that operators of large properties are less likely than others to hold that profit from farming is more important than the environment. Nevertheless, they are more likely than others to oppose additional environmental policy measures. Obviously, constellations of environmental attitudes are more complex than often assumed. The final chapter by *Rukavishnikov* outlines the advantages and limitations of comparative public opinion research on mass ecological attitudes. He shows how important the wording of environmental questions is in cross-national research on environmental concern. He makes a plea for well-constructed and theory-based methodologies also in light of the overall political and practical significance of social studies on trends in environmental attitudes.

1. The Trap of Science and Technology

Pieter Tijmes

Abstract

Attention is paid to the rise of science and technology as a specific way of interpreting reality. It is a remarkable paradox that on the one hand science and technology are very modest within their own domain, but on the other they do not actually tolerate any rival interpretation. That science and technology are revolutionary activities is not an overstatement. Interpretation and change go hand in hand and revolutionary interpretation means revolutionary change. Being morally blind, science and technology nevertheless reconstruct reality. In our world the dominance of scientific and technological culture has been strongly reinforced by economics. The alliance of the homo faber and the homo oeconomicus is not an unproblematic one. This chapter ends in a criticism of technocracy, a call for philosophical research of the meaning of technological context and a plea for ascesis.

1. INTRODUCTION

As everybody knows, Adam Smith regarded steering society to be too difficult and too complicated for ordinary mortals; he thought it wise to leave that job to the Invisible Hand. Nobody will dispute the ideological character of this deistic vision in economics and politics. Diametrically opposed to this, it is nowadays upheld that people ought to take the responsibility for the workings of society themselves and that this task is not too difficult and too complicated because of their access to scientific or technological insights. Playing at the Invisible Hand in this modern way is no less ideological in my opinion, because in this self-satisfied view the paradoxical and ambivalent character of science and technology is overlooked. In this chapter I shall examine this precarious character of science and technology and give the example of environmental policy to illustrate this argument[1].

The main steps in the train of thought are as follows. First of all I shall pay attention to the rise of science and technology as a specific way of interpreting reality. Although it is argued that science and technology are specific ways of interpreting reality without exclusive rights, nevertheless a dominance of scientific and technological culture cannot be denied. From the viewpoint of interpreting reality it is a remarkable paradox, that on the one hand science and technology are very modest within their own domain, but on the other they do not actually tolerate any rival interpretation. In this section I deal with some implications of the fact that the valid judgments of science and technology refer to a limited domain, the so called 'physical' reality or reduced 'nature'. Another important aspect of scientific knowledge is that science and technology are to be considered as creations without morally inherent norms. This aspect asks for the moral problem that the developments of scientific and technological knowledge are in themselves normless. That science and technology are revolutionary activities, is not an overstatement. Interpretation and change go hand in hand and revolutionary interpretation means revolutionary change. Being morally blind, science and technology nevertheless reconstruct reality. In our modern world the dominance of scientific and technological culture has been strongly reinforced by economics. Economics is also a successful agent of interpreting and changing the modern world in line with science and technology. The alliance of the homo faber and the homo oeconomicus is not an unproblematic one.

Finally, I shall focus on the correcting interventions in the environment according to the standards of science and technology. This is a concrete example of their penetrating role in conjunction with economics into society. The discussion of this example ends in a criticism of technocracy, a call for philosophical research of the meaning of technological context and a plea for ascesis.

2. RIVAL INTERPRETATION

The rise of science has been accompanied with much controversies, because the legitimacy of the scientific assumption, for example the homogenizing of nature, has been disputed. That science has gained independence, has been described as a form of secularization. That means that the domain of modern science has been withdrawn from the patronizing by religion and metaphysics, because the domain has been investigated and understood in terms of its own regularities and patterns without interference of a truth from the outside. In this sense secularization and emancipation are the same, but science did not hold on to this. Modern sciences in the sixteenth century did not only produce modern scientific theories that made the prevailing theories out of date. The new theories implicated a mechanization of world view that could hardly be reconciled with the prevailing religiously inspired world view of the Middle Ages.

The Copernican work, *De Revolutionibus Orbium Caelestium*, made the earth give up its central place to the sun. This facelift of scientific knowledge became contagious for questioning the whole religious and metaphysical building. The plurality of reality's interpretations led to rivalries and conflicts, whereas the triumph of scientific interpretation did not remain limited to its own domain, but also spread to other domains. Thus scientific discoveries and developments have often unfolded themselves as subversive powers in the presence of religious ideas, even if these ideas flourished on their own and not in the scientific domain. Under the pressure of this power, religious interpretation had to provide a new make-up in order to survive. In the process of defending its own concern, religion tried to immunize its new point of view against science. Simplifying the matter a little one might say that it turned out to be a new difference between faith and knowledge, on the understanding that the interpretation of reality as creation has nothing to do with the origin of the universe. This history of the struggle between religious and scientific world views ranged over the centuries up to the recent era of Marx, Darwin and Freud. The thought that science destroys religious world views might have become unfamiliar to us, because we have short memories. Maybe we can talk of an agreement for cessation of hostilities. That is to say, we came to terms on the fact, that every reflection has got its own perspective on reality: science and religion with their own outlook on life are allowed to live in peace with each other. What in the heat of the rivalry could not be combined by the representatives of science and religion, seems to be possible now. And it looks as if the animosity between science and religion belongs to the past, thanks to the partition of perspectives. Not only science, but also religion now have become more modest. Nobody wants to play first fiddle since he is aware of the danger of extrapolating knowledge obtained as a result of a methodological construction. But this does not mean that the different interpretations of reality will not come cross each other. The plurality of interpretations might always be a reason to dig up the hatchet again, because interpretation is not only a private way of *looking* at reality, but also gives rise to interfering *actions* of people. In short, on the one hand looking back to the so called peace agreement we immediately recognize that talking of religion as a *perspective* made a big hole in the religious budget compared with times when religion had exclusive rights of representation. On the other hand, one may argue that the dominance of religious culture has been replaced by the dominance of scientific and technological culture, that today only respects some religious niche. In other words: those who speak of complementary perspectives, as I did above, might overlook the real balance of power.

3. AMBIGUITY OF SCIENCE

From the viewpoint of interpretation science shows a remarkable paradox. On the one hand, science is modesty itself trying to reveal the truth and nothing but the truth of reality. This modesty is the power of science and gives scientific truth a high authority. In politics, for instance, science based arguments are the best possible to convince the opponent. That these arguments are not always decisive in the political debate, does not necessarily play down the authority of science. On the other hand, science is an imperialist not tolerating any rivals. It pretends to be exclusively in the service of the truth, but in the course of its acting science has made it abundantly clear that it appreciates only its own truth, because it actually reduces the plurality of interpretations. This imperialistic character does not only imply a reduction of interpretations, but the cultural plurality is also reduced to uniformity. Regardless of what you are looking at: television programmes, ready-made clothes, new estate of houses, architecture, it is all the same all over the world. Taken with a pinch of salt, the Hilton Hotel and the Holiday Inn are the symbols of planetary *uniformity* of culture. This ambiguity of science - humble and tyrannical at the same time - needs an explanation.

The truth, or rather the valid judgments of science, refer mostly to a limited domain. These judgments do not cover the whole of reality, but refer only to a part of it. Science isolates its own domain out of reality and makes its statements about it. This domain is called the object of science. It is not found in reality, but it must be constructed by science itself; this construction means an isolation of a part of reality. In other words, nature as an object of science is different from nature as given in daily experience. Nature as an object of science is the physical, homogenized nature and not nature of the nature lover. Science limits its domain of validity and it tries to formulate the functional relationships between the phenomena. For example, if water freezes the volume increases. In applying physical knowledge to our world of daily life, the scientific isolation of the object has to be undone. When water in the radiator freezes, the heating pipes might burst. In that case, what is correct in theory might turn out to be valid in practice or not. Abolishing the limitation of the 'physical' reality may have consequences which have not been anticipated in theory or which have not been accounted for methodologically. The purer or more theoretical scientific knowledge is, the more the knowledge runs the risk of irrelevance or rather uselessness, when introduced in the world of daily life. What is valid for the isolated domain, the object of science, may be worthless, if the relations of the constructed domain are again put into the world of daily life. A bridge may collapse, even if the calculations of the engineer on the drawing board were correct, because in an unforeseen way difficulties arose; afterwards these unanticipated circumstances might be explained scientifically. In this way science is always right in advance or later on. Application of scientific knowledge is also very complicated for another reason: science itself does not always speak with one voice. As for the acceptable concentration of toxic chemicals toxicologists supply more disturbing messages than epidemiologists. The former base their messages on toxic effects on organisms, and the latter are used to look at the toxic effects on the population as statistical entity. The tenor of the argument is often contradictory. Where the one warns against toxic material, the other holds the conviction that one can live very well with a little bit of it. These opposite advices refer to different parts of the whole of reality.

4. SCIENCE WITHOUT NORMS

Science and technology construct a homogenized nature that can be mathematically managed. Physical nature is available for everything to be done, on the condition that its laws are obeyed. This is the only clause. In this sense modern science and technology are to be considered as creations without inherent norms. The developments of scientific and technological knowledge are in themselves normless and do not know any intrinsic measure or limitation. Knowledge on scientific and technological interventions only refers to what is possible. Embryo's are, as it were, to be bred as many days, weeks or months as you want, energy is to be generated as much as you deem fit, bombs are to be exploded, as you like it. The fact that science and technology have no measure or limitation, means that we do not know on scientific or technological grounds, whether we follow a good or a bad track. Here lies the origin of science's ambivalence. Good as well as bad results are often interconnected: you cannot have the one without being confronted with the other. It is not so easy to follow the biblical advice "examine all things and keep what is good". On top of this, good and bad results are often unintentional and unlooked-for. In physical reality there is moral neutrality, but the scientific and technical products appear in the world of daily life, where their effects are not at all morally neutral. That is the reason why many technical products, for example the car, the computer, the nuclear power station, the TV as well as technological procedures, for example abortion, euthanasia and so on, get their ambivalent character. Moving by car from the one point to the other is going fast, but because of the car's democratization it cannot be considered from the viewpoint of moving only. The power plant may produce cheap and clean energy from the economic and environmental point of view, but it is also to be concerned as a bomb for security reasons and as a source of non-disposable radio-active material and so on. Much is possible in physical nature, but not everything is useful in the context of the world of daily life with its moral discourse. This world is the stage of the ambivalence of science and technology. In other words, the changing contexts of the two worlds, the world of technology and the world of daily life, let the technological artifacts appear in a different light. The changing of context makes it impossible to defend the moral neutrality of science and technology in our life. Moreover, also in the world of daily life these technological products must be evaluated within different frameworks. Unambiguous statements are not to be expected.

5. THE RECONSTRUCTIVE POWER OF SCIENCE

In interpreting reality by science, a scientific net is cast over it and it looks as if reality adapts to that net. That could mean that the world itself - and not only our interpretation - has changed since the scientific revolution. Two examples out of the medical field will illustrate how interpretation and change of reality can go hand in hand. In the first half of the seventeenth century, the discovery of blood circulation by Harvey met much resistance. The concept of blood circulation redefined the human body. To accommodate the circulation the body interpreted by tradition must be recast as a functional system of filters, conduits, valves and pumps. For people in Harvey's time their body was alienated in a certain sense. One could argue that the ambivalence of science consists in the fact, that we have bought treatment of heart attacks with this alienation. This example has probably little appeal for us, because we as modern men do not understand and can hardly retrieve the alienation, whereas we are quite happy with the extension of the medical package. Nevertheless my point is here that the discovery of new medical facts means that we have got a new body. Thus science does not only change our interpretation, it also changes the world. A second example. The prenatal care of the pregnant woman has drastically changed in the last century. The responsibility for pregnancy has been transferred to medically professionalised male hands. This medical profession has discovered and visualized the foetus of the pregnant woman. In the past women were said to be expecting; because of the prenatal care all expecting will have gone out. The woman knows already the gender of her child she will bring into the world, she has excluded the possibility of Down's syndrome etcetera and in the future she will be kept informed of the medical prospects of her foetus. And she will act according to this information. What is good and bad in prevention of oligophrenic, of mentally handicapped, of undesired or less desired life? In any case as a result of the woman's inserting into the scientific framework, the social connotations of expecting have changed. The result is a new relation to her body and a new perception of her inner self. Both of them will have been changed. It is by her free submission to the medical and professional care that the experience of her own reality will drastically change. Is it a clear advantage that the future generations cannot withdraw themselves from these revolutions? This affords a glance of the scenes of the scientific and technological culture. Above I wrote that science and technology do not know any ethical norms and that only in the context of the world of daily life the products and procedures ask for ethical norms. But it looks as if in the established prenatal care the scientific culture has got the exclusive rights of interpretation of reality and implicitly also the right to tell what is to be done. What is medically possible, for example preventing mongolism, becomes an imperative. This example sharpens the insight into the fact that the undisturbed or rather the anarchic development of science and technology permanently generates and imposes new facts, new interpretations of reality and new invitations how to deal with them without being asked for. (I call the development of science and technology anarchic because of the unforseen possibilities of combinations of technological products and procedures. Nobody can predict their development, because it has not an inner necessity.) The question whether modern and scientific interpretations are better, more preferable or richer is beyond the scope of this culture and remains of course unanswered. More generally formulated, science and technology are deeply penetrating and reconstructing our daily life. Modern meanings and interpretations supersede and blot out the old ones. Consideration is nevertheless not out of place whether we do appreciate these substitutions. In my opinion it is of utmost importance to do research into experiences and interpretations of reality preceding our modern time and which our scientific culture has vigorously and energetically destroyed. The plea for research like this has nothing to do with nostalgia, but is based on the consideration that it is not without importance to know the cultural expenses of our modern way of life in contrast with the past. Traditional articulations are not to be used as models; of course they can not be normative. Their potential surplus value, if any, has to be demonstrated.

6. ECONOMICS AND THE WORLD OF DAILY LIFE

So far we discussed three aspects of the modern twin brother, science and technology: the nature of their knowledge, their changing power of reality and their cultural influence on the way people think and behave. The result was that the traditional world view has disappeared. Looking at economics we may discover that the same phenomena are even more vigorous. Some centuries ago economic thinking made its own contribution to the destruction of the traditional world view in Europe and America. The relation between economic knowledge and economic reality is most intimate. On the one hand, the assumption of modern economics is a modern world and it is only that world that economics can convincingly interpret. On the other, it is remarkable that this modern world on its turn adapts itself to the economic interpretation: cutback in taxes raises patterns of spending; growth of G.N.P. is accompanied by improvement of living standards and so on. But although these formulas have a certain plausibility in our own economic society they are not very successful in traditional societies that do not produce for markets, but only for their own consumption (subsistence-economy). In the economic frame of reference one can only fight poverty by means of economic growth and see well being only arising from commercial intercourse and transactions. That is the reason that the famous Brundtland report calls for participation of the underdeveloped countries in the world market with its free circulation of capital and commodities. This well meant call has its origin in our modern world view and sees the problems through economic spectacles, although these traditional societies are not economic ones in our way. The modern economic knowledge does not match the traditional domain in this case. The Brundtland report overlooks the fact that this way of thinking and acting leads to an increased vulnerability of these populations. In the report this vulnerability does not seem to be a question worth considering. The modern variant of the rationality of economic thinking adorns itself with the term 'sustainable development' coined by the Brundtland commission. By defining sustainable development as a 'development that meets the needs of the present generation without compromising the ability of future generations to meet their own needs' the report remains within the economic logic, where needs are infinite and means are limited. Those who do not dare to indicate a fixed ceiling for the present needs of the rich countries and only argue in favour of selective economic growth with the intention of paying the aspirations of the poor countries out of it, are a victim of modern economics with its assumptions. The economic orientation might be even more expansive and penetrating than the scientific one. The triumph of economics is reflected in the way we speak about our world and fellow beings in an economic idiom. Economic thinking and speaking do not only refer to economic phenomena, but have become a world view, whose articles of faith everybody seems to confess. For example: time is money; energy is invested in personal relations; education is a depth investment in youth, culture politics is a question of supply and demand; and so on. We have become people having an economically twisted mind.

In short, both economics and science and technology, reinforcing each other, daily demonstrate their transforming power of reality, that undermines patterns of culture by imposing a new uniformed interpretation. This process has taken place in our own history of the Western world and takes place in those countries which have been baptized as underdeveloped countries after World War II. In practising economics, science and technology, specific possibilities for living are chosen and others are vigorously excluded. It is advisable we pose the problem whether the success of the uniform pattern is an indication of its wealth. Another point is to be added. Problems we are now confronted with on a world scale such as armament (1), hunger (2), environment pollution (3), psychic uprootedness and alienation (4), have been created to a large extent or at least aggravated by science and technology. Of course, this statement is not a secret plea for getting rid of them, but only emphasizes their ambivalent character and criticises the technological solutions to these problems. (1) Everybody agrees that nuclear armament has given evidence of a typically technological approach for security. (2) In literature on development aid you are confronted with the fact that hunger is most difficult to be fought in those places, where the Western culture has torn up the traditional bonds of solidarity. (3) Environment degradation is the result of our aim of safeguarding - and if possible of increasing - the attainments of affluence on a scientific base. (4) It becomes more difficult to put the blame of alienation on anything special. Nevertheless one can hold the conviction that these crises get rather worse by science and technology than that they are being resolved.

7. ENVIRONMENT CORRECTIONS

Finally, I shall discuss a concrete example of the penetrating role of science and technology in conjunction with economics into society and raise the question whether they are our partners or enemies in the struggle for preserving nature. Our attention will now be focused on the correcting interventions in the environment according to the standards of science and technology. Two levels are to be distinguished in environmental affairs. The first level is that of the scientific and technological interventions. Phenomena as environmental pollution are the result of scientific regularities, they are unintended side-effects of technological procedures. Because of the perceived technological character of these side-effects, one can try to find technological solutions for the problems. These solutions, however, are not only technical, just by calling them such is a way of reducing their nature. In any case, the detrimental side-effects show us that reality always remains more complex than our scientific models. After three centuries of scientific and technological intercourse with nature we have to acknowledge that nature cannot always be manipulated without dangerous consequences. So far we have removed the problems we were confronted with, only for other problems to take their place. Speaking about regrettable side-effects of our technological solutions is short-sighted. We pretended to give solutions for improvement refusing to face facts and side-effects. What seemed to be established as adequate for the short term, proved to be an accumalation of arrangements clashing with nature in the long term. Starting from the assumption that nature was only the object of science we have forgotten that in a certain sense nature was also a subject that could not always be manipulated and reduced to a homogeneous field for the glory of manipulation by us. In that reduction the fact was overlooked that it gave us our life. Now nature threatens to give up that function and does not correct our exploitations any longer. Life disappears out of the seas, rivers become open sewers, forests die off and nature will be no basis for human life any longer.

The second level to be distinguished in environmental affairs is that of the political interventions. The effects of the reductive treatment of nature are also to be felt on the level of society. So far as the first level of scientific regularities is transparent, it is also possible to try to prevent the damaging effects on the level of issuing of political rules. One can decree that certain acts are forbidden. For instance, all procedures degrading ozone-layers and furthering the green house effects are to be forbidden or strongly reduced by environmental legislation. With science as a shield environmental protection may succeed. But who thinks to make order out of chaos by means of political reactions, runs after the facts. Better late than never, one may say. The pattern of thinking and acting is the same as on the first level. Society is represented as a homogeneous space for intervention by the political engineer. Of course that is technocracy; and although technocracy dealing with social and political problems has not very good testimonials, we are a bit touched by it in the sense we all believe that a rational approach, that means a scientific approach, is the best way of dealing with our problems. Our modern creed is that a problem can only be solved scientifically. Just in the same way as nature resists to the offenses in the long run, so society reacts to these legislative interventions. It also has a dynamic of its own. That can be ignored for a long time, but not for ever. What then has not been taken into account? In society people live with their own desires, ambitions, aspirations, agenda's and so on. Hunting for realisation of their wishes they will consider the legislative interventions as problems to be solved. All according to the rule: what is not forbidden, is allowed. The bureaucracy on the other hand is always ready to close a loophole in the law. The result is that rules after rules are formulated, and the end of the story is that those who think up these rules do not know the whole of rules any more. At last he makes the proposal for deregulation, as if the rules were pointless. This way of dealing with problems explains the compactness of legislation, and the bureaucratic arbitrariness and capriciousness. My point is here that modern bureaucracy can be considered as a pinnacle of the engineer's mentality.

The short term solution of intervention into nature has its exact parallel with the political and bureaucratic intervention into society. That corrections are rapidly needed, gives evidence to the fact that in origin the measures were not very good. Holding the Popperian conviction that we may learn from our errors often means that we incorrigibly go on implementing corrections after corrections. We do not see that the path of correction has no end. Our scientific approach, pinnacle of rationality, does not observe the irrationality of our endeavour. Is it not time to pose the question whether our effort to steer society on a scientific base does not make society more unsteerable?

Keeping ourselves out of range

We have to tremble for our interventions in nature to improve life conditions. Environmental degradation is not a technological problem to be solved by technological corrections, reparations, compensations, etc. Neither is it a social and political problem to be solved by social and political measures. Even the advocated two way approach - technological and political - will not lead to solutions acceptable in the long term. Experiencing the limits of scientific and political manipulations we have to discover ourselves to be the problem. Why do we keep ourselves out of range? The problem of the environment is the result of our lifestyle and our level of aspirations. Who tries to solve this problem without considering modern forms of ascesis, forms of voluntarily resigning high consumption levels, looks for solutions in the wrong direction. We actually give the impression of not renouncing the economic assumptions of infinite needs and limited means. In modern economics this strategy of infinite desire for affluence is taken seriously without paying attention to the irrationalities attached to it. It is a strategy that hunts for material fulfilments without measure and satisfaction. In spite of our efforts to fight scarcity during the last centuries the experience of it has nonetheless remained, because scarcity is not a characteristic of nature but of human relations (Dumouchel). Here lies the root of scarcity: comparing with others we often feel being short of things the other has got. It is this comparison - as universal phenomenon it is typically modern - that makes us unhappy or jealous and charges us with feelings that we are underexposed, undergifted, underpaid, and so on. It is important to take notice of the fact, that the hunting for improvement of affluence does not admit a stable level. Affluence is not a fixed but a receding goal since it depends upon our infinite craving. We will never be happy with any conceivable level of goods and services. A plea for ascesis is a plea to indicate a ceiling for the present needs of the rich countries. Above I wrote that environmental pollution has been generated or sharpened to a large extent by science and technology, but the motor of the pollution may be located in the institutionalization of the infinite desire, in the economy.

Contribution of philosophy

Let me try to reformulate my criticism of the implicit technocracy in environmental policy. What is called reality on the first (technical) and the second (social) level, has been objectified as a challenge for modelling by the modern *homo faber*; it is his profession to face challenges. The assumption of this objectivation is that the technological approach only needs to be refined in order to avoid the so called detrimental side-effects. Under modern conditions homo faber and technocrat do not only belong together but coincide. If the corrective measures miss the mark, they only have to be made more sophisticated. My philosophical point of view is that the concept of the homo faber - and in the wake of this concept, his radiation in practice as a result of the fact people think about themselves in terms of this concept - has to be questioned, because this concept is also a reduction of mankind. What is wrong with it? What has been forgotten, if you only think of mankind as homo faber? The concern of this philosophical reflection is not refinement or perfection of technology, the homo faber's duty, but rather its meaning. A plea for refinement takes the prevailing context of technology as given. The philosopher who concerns himself about meaning asks for the context itself.

8.   IN RETROSPECT

In this chapter I first paid attention to the methodological limitation of scientific knowledge. This limitation sharpens the eye for the science-based recommendations which are sometimes contradictory and often too limited. Second, I focused on the inherently ambivalent and revolutionary character of the interpretation of science. The assumption that scientific and economic knowledge supplies the ingredients for our modern world view relieving the guard of religion, has been criticized as getting out of the frying pan into the fire. In this relieving of the guard there is no question of an end of ideology, as usually supposed. The environmental policy with its well meant corrections exclusively based on science and technology is an example of the technocratic approach. The result is that not all the facts relevant in view of a possible solution are taken into account. In this context the plea for ascesis - an interesting historical possibility, although under different conditions - should not be interpreted as repeating the obsolete, but as a call for reflection on what we really want.

REFERENCES

Berg, J.H. van den. (1960). *Het Menselijk Lichaam, I/II*. Nijkerk: G.F. Callenbach N.V.
Berg, J.H. van den. (1970). *Things: Four Metabletic Reflections*. Pittsburgh.
Dijksterhuis, E.J. (1950). *De Mechanisering van het Wereldbeeld*. Amsterdam: J.M.Meulenhoff.
Illich, I. (1986). *H_2O and the Waters of Forgetfulness*. London: Marion Boyars Publishers.
Pot, J. H. (1984). *Die Bewertung des technischen Fortschritts*. Assen: Van Gorcum.
Weizsäcker, C. F. von. (1981). *Der bedrohte Frieden*. München/Wien: Carl Hanser Verlag.
Weizsäcker, C. F. von. (1980). *Der Garten des Menschlichen*. Frankfurt am Main: Fischer Taschenbuch.
Weizsäcker, C. F. von. (1988). *Bewußtseinswandel*. München/Wien: Carl Hanser Verlag.

NOTES

1. In politics, science and economy are mostly considered as interpretations of reality and it sounds rational to interpret the world before changing it, despite Marx. Nevertheless it will turn out, that the relation of interpretation and change is more elaborate, for the very reason that an interpretation is an open or implicit invitation to specific actions, or even stronger an interpretation amounts to a change.

2. Capitalism, Nature and Modernity.
A Realist Perspective

Peter Dickens

'The idea of one basis for life and another for science is from the outset a lie. Natural science will in time subsume the science of man just as the science of man will subsume natural science: there will be one science' (Marx, 1844).

Once, it seems, we knew what to do. Until the early modern period, knowing who we were and in what practice we were engaged, told us all we needed to know about what we ought to do (Szerszynski, 1995).

Abstract

*Contemporary societies are characterised by highly complex divisions of labour. Technical divisions of labour within the workplace and broader social divisions of labour all result in a highly fragmented understanding of society-nature relations. Abstract theory is separated from practical knowledge. The natural and physical sciences are separated from the social sciences. The technical division of labour ensures fragmentation even between those workers in the same industrial enterprise. All this leads to a particular kind of alienation; one emphasised by Marx 150 years ago but now largely forgotten. This is alienation resulting from a failure of people to **understand** their relations with nature. Yet as human society increasingly impinges on the natural world such divisions of labour become more problematic. As a first step towards remedying the situation, this chapter proposes a new alliance between the social and life sciences, one which develops an almost lost approach to understanding the relations between organisms and their environment.*

1. INTRODUCTION

What type of theory is 'best' for the understanding of people-environment relations? Is a completely new type of understanding required or can we build on existing perspectives? This chapter takes the latter view and adopts one aspect of Marxism as a suitably robust basis for the construction of a sociology of people-environment relations. But the argument parts company with most of those contemporary Marxist commentators on the environment. The chapter argues that important parts of Marx's message have not so far been recognised. In his earliest writings in particular, he recognised 'man' as a natural being; one that was being alienated by the social and environmental context which 'he' was creating. Thus people are seen as possessed of creative and capacities. They also need an association with nature to develop as human beings. Capitalism, Marx argued systematically alienates people from nature, from one another and from the products of their work. But people can to some extent regain a sense of restored personal identity. This can be more easily achieved outside employment. The purchasing of commodities has a particular importance, although other activities (the many forms of 'voluntary action' people undertake is a case in point) are where a degree of fulfillment can be regained. But it is primarily in the purchasing of commodities where people can gain a sense of self-fulfillment. In short, the other side to the coin of alienation is that of commodity-fetishism. These aspects of Marx are familiar. Furthermore, their application to environmental analysis is now becoming more commonplace (see, for example, Grundmann, 1991; Dickens, 1992; Benton, 1994; Pepper, 1993; O'Connor, 1995). It now seems surprising that social theory neglected the environment. And Marx, with his core recognition that all societies must transform, modify or alter nature, is an obvious source to turn to in trying to develop a 'green' social theory.

2. CONTEMPORARY ECO-MARXISM: CAPITALISM AS THE WRECKER OF NATURE?

The debate so far has largely revolved over whether Marx really is a sound basis for an ecologically-sound analysis of people-nature relations. Eckersley (1992) argues strongly, for example, that it is not. She says that 'humanist eco-Marxism' posits an inverse relationship between alienation and the subjugation of nature. True human freedom, in other words, lies in the the human subject freeing themselves from a dependence on external nature and its limitations. They must in effect 'master' it if they are to be free. If Eckersley is right, therefore, Marx (and his understanding of how people are to be emancipated) lies firmly within the human-centred enlightenment tradition. Nature is simply there for the gratification of human beings and this is a poor start to environmentally-benign social theory and political practice. Against this, however, O'Neill (1993, 1994) and Hayward (1995 forthcoming) argue that there is another way of reading Marx which recommends itself much more warmly to contemporary environmentalism. Parts of Marx (and it is not really surprising that Marx's early notebooks were sometimes contradictory) are open to what O'Neill calls 'an ecologically benign interpretation'. To 'master' or to 'humanise' nature does not necessarily mean to physically conquer it. Such mastery can consist (as both Engels and Marx suggested) of *understanding* its properties and qualities. This partly means identifying with the natural world in an aesthetic sense (and in this sense developing human subjects' alienated aesthetic capacities) but also gaining a scientific appreciation of its underlying mechanisms and tendencies. Whether or not this is the view which Marx originally adopted, it is this latter sense of 'humanising' or 'mastering' to which this paper subscribes. But this brings us to yet another problem and, again following Marx's fragmentary notes, another source of alienation from nature for human beings. It is one which has received very little attention from the literature. This is one whereby humans fail to understand their relationship with the environment because they have a wholly inappropriate set of concepts for creating such an understanding. One of Marx's most important early insights was that a re-alignment or fusion between the sciences would have to take place, one which recognised the extent to which 'man' was increasingly humanising nature. Bearing this in mind, while Marxism has found increasing acceptance in the contemporary sociology of the environment, this central message about the division of labour as a central cause of alienation of people from their environment has been largely lost. Perhaps this is because sociology and professional sociologists (including those influenced by Marx) are not keen to lose or relax the disciplinary boundaries acquired during the late 19th and early 20th centuries.

Manicas (1987) has documented with great care the process whereby these intellectual divisions of labour came to be in place. They were the result of a very particular historical conjunctures. Late capitalism had thrown up a series of 'social problems' especially in Germany, Britain, and the U.S.A. Groups of university-based professionals presented themselves as capable of addressing and solving these problems as detached and often apolitical 'experts'. Furthermore, they latched on to and adopted a stereotypical view of how scientists in the physical and natural sciences worked, a view which placed particular stress on its supposed rationality, objectivity and openness to falsification. The key problem with all this was not only that the model of science was misleading but that science itself was left out of these new ways of constructing understanding. Peoples' relationship with their environment were not seen as part of the very social problem they were trying to understand. The nature of knowledge about society and the environment seems, therefore, to have escaped most 'Marxian' approaches to the environment. It has escaped other forms of sociology to an even greater extent. Yet such a gap is surely inadmissible. Knowledge is a central part of how human societies labour on nature. The concepts we as humans use to understand nature all have real material effects on the natural world. At the same time, the way we physically interact with nature to produce the things we want changes our conceptions of that same nature. As Marx argued, in changing nature we change ourselves. Thus the environmental question and a host of other related issues are pressing for an urgent engagement between such hitherto separate areas of thought as biology and the social sciences. As Dunlap (1980) has long been arguing, the type of social science that developed to deal with a particular set of circumstances during the 19th century now seem increasingly inadequate. A new set of sciences or intellectual division of labour is needed, one which can deal with a pressing new range of problems. And these are of course very explicitly 'political'. These issues include the environment, human health, animal rights and even gender. As Benton (1991a) has asked, how much longer can sociology insist on a science which does not recognise the insights of biology and even of the physical sciences? How much longer can social constructionism be used as a substitute for a new science which starts to link social theory into the life and physical sciences?

Marxism, or at least a version of Marxism, has indeed made considerable strides in helping to link social structure and social change to the physical and organic environment. But it has not yet recognised the logic of Marx's argument that knowledge itself must change to reflect human beings' impact on nature. This perspective has indeed been central in introducing into social theory an understanding of society as both the product of our interaction with nature and itself while also affecting society itself. In the recent literature perhaps most important is the attempt by O'Connor to revive Marx's 'Second Contradiction'(O'Connor, 1988)[1]. Conventional Marxist theory of course stresses the contradiction between the forces of production and the relations of production. Capitalist technology enables goods to be produced which cannot be sold. Their value cannot be realised as a surplus. Social and political crisis ensues, according to this 'First Contradiction', with the potential for socialist revolution and social transformation led by the working class. All this is relatively familiar. But, as O'Connor points out, there is a second type of contradiction which has remained relatively unrecognised by contemporary Marxism. This is the contradiction between capitalist social relations and productive forces and the *conditions* of capitalist production. These latter are what Marx called 'natural wealth in means of subsistence.' Today we give much more stress to these conditions. They include ecosystems under threat, the thinning of the ozone layer and pollution of many kinds. Other types of 'condition' were recognised by Marx. They include workers' capacity to work, or labour-power. Thus we are here talking of human beings' health and, in the present era, threats to their physical-cum-mental well-being resulting from the ways contemporary society changes nature. Taking these conditions on board must entail recognition of this second type of contradiction, or another way in which societies transform themselves through transforming nature. Although the evidence is by no means clear, it is at least possible that capital is (in some instances quite literally) undermining the potential for its own growth and development. It is in effect wrecking the external conditions necessary for its survival while ruining the workers' internal nature, their capacity to work As O'Connor puts it: The main question - does capital create its own barriers or limits by destroying its own production conditions? needs to be asked in terms of specific use values, as well as exchange values (p. 25). This 'second contradiction' has also been subject to considerable amplification and empirical analysis (see for example Martinez Allier, 1993; Ravaioli, 1993; O'Connor (ed.), 1995). All of these reach much the same conclusion. This is that capitalist growth (and growth by the competing East European economies) has systematically ravaged the environment, resulting in increasing pollution, destroyed forests, desertification, changing climates, the weakening of the ozone layer, loss of biodiversity and a generally worsening quality of life.

This chapter does not part company with these analyses. But, again, it sees them as offering only a partial understanding. Such a picture is necessary but not sufficient. There are important features of Marx's original account which are necessary for a deeper understanding. Especially important, as outlined earlier, is the division of labour. This applies to the technical division of labour and, still more importantly, to the social division of labour. The technical division of labour usually refers of course to the division of labour within a firm. When it comes to an understanding of how capitalist industry can alienate workers and wreck the non-human world it is important to remember that *all* large-scale labour-processes will need some form of control and direction. This applies whatever the particular type of modern society exists or which may be created in the future. And, Sayer and Walker put it: 'Command over large organisations, research and marketing activities, or the circulation of money all imply very different levels of control over social production. All this is essentially independent of the class relations in modern industrial economies' (1992, p. 18). It is therefore difficult to imagine the absence of an extensive division of labour in any future form of modern society with, more significantly still, hierarchical relations of management and supervision of labour. It seems to be, so to say, part of the modernity's 'package.' It certainly cannot be avoided by appeal to mythical socialist futures in which there is no longer any division of labour and the management by certain classes of other subordinate classes. We might parenthetically mention here that is also difficult to imagine forms of technology in modern society (capitalist or socialist) which offer more direct or sensuous relationships with nature (Grundmann, 1991). Confronting the impact of the division on peoples' understanding of their environment means *inter alia* dealing with what Wainwright (1994) calls 'practical knowledge'. As feminist scholarship in particular has stressed, experience does not simply produce facts which confirm or falsify laws. It can also provide insights into these same underlying structures and relationships. Indeed, feelings of this kind may be the origins of new assessments of abstract knowledge.

'Experience, rather than simply yielding facts which confirm or falsify general laws, provides clues to underlying structures and relationships which are not observable other than through the particular phenomena or events that they produce. The precise character of such structures can only be understood by paying attention to the details of experience of the events and phenomena that they generate, its variations as well as its recurrences. Moreover, feelings can be signs of an inadequacy in an influential interpretation of experience' (Wainwright, 1994:7).

And yet the importance of practical experience and its role in linking to and interrogating theoretical knowledge has been consistently down-played by modern science, including social science. The fact that the division of labour (and especially that between mental and manual work) has been so ignored is all the more surprising for the fact that Marx himself increasingly saw such as the main obstacle to human emancipation (Rattansi, 1982).

The rigid division of labour is not simply protective of academic disciplines. It is again of course a fundamentally political question. As Wainwright again puts it 'the democracy of doing' needs to be placed alongside 'the democracy of deciding'. The latter is the much more familiar formal democratic processes in the political sphere but a further form of democracy surely lies through recognising different sources of knowledge and sharing them between 'experts' and others. The importance of the fragmentation between practical and theoretical knowledge is very significant as regards understanding peoples' relationship with the environment. It means that much more credence needs to be placed on practical understandings of people-environment relations and, furthermore, the incorporation of such understandings into the organisation and management of work. Much of contemporary environmental thinking places, of course, a particular emphasis on local knowledge and on deriving a general and global understanding based on peoples' experience which is necessarily limited to regions or localities. But there seem to be few institutions or mechanisms which allow such forms of knowledge to be linked. Within the detailed division of labour within the workplace, there will always be considerable potential for clashing interests and major problems of coordination, whatever the type of society in question. Remembering the experience of the previously state-socialist societies, the hierarchical management of collective labour by managers entails, for example, the continuing separation of mental from manual labour or 'practical knowledge'. The mental labourers and the managers will usually be able to set the environmental agenda and to be highly influential in establishing environmental strategy. Typically they have the factories in their charge and they will be prime possessors of the knowledge of how their enterprise the physical and natural environment. 'Practical knowledge', and 'the democracy of doing', will again tend to find themselves subordinated or marginalised by managers, but this is the kind of challenge made to the social relations of the workplace by contemporary environmental problems and cannot be ignored. But even this is too simple. Different *managers* within the technical division of labour may, for example, be charged with contradictory tasks. One manager in an advanced socialist or capitalist factory may be charged with meeting production-targets while another is charged with monitoring and controlling pollution. Which priority is to prevail? This kind of dilemma seems to have been a particularly common one in the previously state-socialist societies aiming to produce at all costs, including at the cost of the environment (Tamas, 1992). This only leads to the bigger question concerning the division of labour and the environment. The technical division of labour refers to the division within an enterprise, one in which a single person spends the whole of his or her time carrying out a particular task. The social division of labour in modern societies (whether capitalist or the various types of existing socialism) is of course wider. It refers, for example to the division between labour (or 'practical knowledge') in the home and that at the workplace. Taking this wider view, how do the 'environmental interests' of a home-worker (someone, say, who has particularly responsible for a child's health) match up to the 'environmental interests' of someone in paid work. Are they necessarily compatible? And again, how are these different types of practical knowledge rcognised, not just by the formal political process but as forms of knowledge which are valuable in their own right? Furthermore, and most importantly from the viewpoint of this chapter, the division of labour includes divisions between different types of *mental* labour. An industrial chemist and a teacher of sociology for example, will have very different ideas and often very conflicting views about how the relation between the environment and society. They may have quite clear views of this relation. But there is absolutely no guarantee that their understandings are compatible with one another. Indeed, there is every chance that with their own specialised discourse they will continue to mis-communicate, to talk past each other. Other divisions within the social division of labour may be just as important. Consider, for example, the food chain. Manufacturers of fertilizers, farmers, farm-workers, retail outlets and people who act as 'customers' all have widely different and frequently conflicting types of knowledge (particularly the intellectual knowledge associated with the producers as against the primarily practical knowledge of the consumers) as to their relation to the environment and what may be appropriately 'green' actions. And again, it is difficult to imagine these divisions and differences fading away with the rise of another type of 'environmentally friendly' modern society. As things currently stand, any successor to capitalism will presumably also be based on complex divisions of labour and on the present (fragmented) state of knowledge. It will be subject to the same problems of the division of labour, that between practical and theoretical knowledge as well as that between different *types* of mental and practical knowledge. Consider too the environmental and geographical roots to the division of labour and occupational specialisation. Marx himself gave considerable attention to this and it is indeed surprising that contemporary environmentalists have not given consideration to the matter. In *Capital* Marx argued that spatial variation in climate and vegetation were the origins of the division of labour and exchange. Each place had its 'special advantages' for different types of economic development. And as Marx put it:

'It is the differentiation of the soil, the variety of its natural products,the changes of the seasons, which form the physical basis for the social division of labour' (quoted in Rattansi op. cit. p. 168).

Even more startlingly, Marx went so far as to argue that the absence of fertility is vital to social and economic development. Localities and countries which are naturally well-endowed are least likely to generate people into developing their own nature as well as external nature.

'Where nature is too lavish...she does not impose upon him any necessity to develop himself. It is not the tropics with their luxuriant vegetation, but the temperate zone, that is the mother-country of capital' (quoted in Rattansi op. cit. p. 168).

Thus Marx is drawing attention here to environmental conditions and their effects on the division of labour and economic development. Arguably much of the recent history of economic development has revolved around denying these limits or obliterating the environmental differences between places. Modern transport technology means, for example, that even if the environment of one country leads to the production of a particular kind of commodity, this can be readily transferred to different points of the globe for consumption. Advances in the genetic engineering of seeds mean that different parts of the physical environment are now able to produce the same commodity. But, at the same time, *spatial* divisions of labour, those based on the continuing natural characteristics of regions and countries, must still operate as limits. The possibilities for emancipation and environmental consciousness within each of these regions and countries are limited and created by spatial divisions of labour. Marx's main concern, as outlined earlier, was with the division between what he called mental and manual labour or intellectual and manual tasks. It was increasingly here that he saw the prime social struggle, capital appropriating scientific knowledge while labour being transformed into mindless drudgery and denied participation in decision-making. But, as his plea for 'one science' clearly suggests, another route to emancipation lay through links between different types of mental or intellectual labour. Each branch of mental labour (for example, biology, physics and the social sciences) has since Marx's day made huge and important developments in its own right. But it has continued to do so at the expense of making connections with neighbouring insights. Outhwaite (1979), following a review of sociology's historical origins, concludes that it is:

'a rather insubstantial discipline with little theoretical ballast of its own, tending, as Marx said of philosophy, to share the illusion of the epoch' (p. 292). If sociology is missing a central 'theoretical ballast' there seems no reason why it should not borrow some from neighbouring disciplines. Perhaps such sharing could turn out to be life-saving for all concerned, including sociologists. How can such sharing take place?

3. THE BIOLOGY OF ORGANISMS AND ORGANISM-ENVIRONMENT RELATIONS: RECOVERING A LOST TRADITION

How would we go about constructing this 'one science'? Less ambitiously, how might we start making links between *neighbouring* sciences? What intellectual resources would we draw on? As a first step in creating Marx's 'one science' this chapter turns to the natural sciences. Darwin's theories have been subject to what Mayr (1991) calls 'one long argument'. But the variety of approaches in the natural sciences has been largely over-looked in recent years. This has been largely a result of the rise of neo-Darwinism. Basing their views on Watson and Crick's discovery of the molecular structure of D.N.A., neo-Darwinists insist that the attributes and behaviour of organisms can be best understood by concentrating on the organisms' *parts*. As many authors are now pointing out (see, for example, Benton, 1991a) this is a very crude type of reductionism. Why should the complexity of an organism and its relation with the environment be reduced to its molecular components? The arguments against sociobiology and reductionist biology and now well-rehearsed and it is not proposed to enter into these debates here (for perhaps the clearest statement see Lewontin, 1993). The argument here is that the older traditions are now being recovered. Furthermore, these traditions are much more in line with Marx's idea of constructing a single science, one linking the natural and social sciences. One theme in this alternative tradition which is very much in line with contemporary 'green' thinking concerns the close, dialectical relationship between organism and environment. A number of biologists could be taken as examples here. They include W.B. Cannon, L.J. Henderson, C.S. Sherington and J.S. Haldane (see Benton, 1994). To illustrate some of the themes they were addressing during the early part of this century I will concentrate here on Haldane (he is quoted at some length to stress the perhaps surprising connections between this 'old' way of thinking and contemporary environmental analysis). In 1913, Haldane was arguing that: 'the relation of the living organism to its environment is no less peculiar and specific than the relationship of the internal parts and activities of the organism itself. Between organism and environment a constant active exchange is going on. But this exchange, in so far as it has any physiological significance, is always determined in relation to the rest of the living activity of the organism' (p. 79). Four years later he was making what was in effect a realist argument in biology, one which envisaged organisms as containing latent structures and tendencies. And, still in line with a realist ontology, he saw these as being realised or 'woken up' by their environments. None of the innumerable structures special to the adult organism are present in the developing ovum; but as if guided by stimuli which awaken memories of its parents and ancestors, it builds up the adult structures and activities by degrees, often event the finest nuances in the character of either parent. In a living organism the past lives on in the present, and the stored adaptations of the race live on from generation to generation, waking up into response when the appropriate stimulus comes, just as conscious memory is awakened (p. 98). This led him on to argue that there is 'prime reality' to an organism. This he saw as being constituted both by an organism's own internal powers and by its relationship with the environment.

We are inevitably forced to the conclusion that the life of an organism, including its relations to internal and external environment, is something of prime reality, since it persists actively as a whole, and moreover tends to do so in more and more detail with enlarging experience, so that life is a true development. What persists is neither a mere definitely bounded physical structure nor the activity of such a structure. There is no sharp line of demarcation between a living organism and its environment. The persistence of the internal environment and its activities is, in fact, as evident as that of the more central parts of an organism; and a similar persistence, becoming less and less detailed, extends outwards into the external environment. An organism and its environment are one, just as the parts and activities of the organism are one, in the sense that though we can distinguish them we cannot separate them unaltered, and consequently cannot understand or investigate one apart from the rest. It is literally true of life, and no mere metaphor, that the whole is in each of the parts (p. 98-99). Haldane was to develop his point about the between organism and environment in more dramatic form in 1935, insisting that the relationship between organism and environment should more accurately be seen as the interpenetration of organisms. Biology, he argued, is at fault insofar as it envisages a universe of individual lives, with each regarded separately. Rather, the fact that the life of an organism extends over its environment implies that the lives of different organisms, although they are distinguishable, enter into each other's lives. There is no spatial separation between the lives of different organisms, just as there is no spatial separation within the life of an organism. But the part played by the life of one organism in the life of a second is, from a mere biological standpoint, only intelligible as entering into the life of the second (p. 64). Latterly, this more holistic way of envisaging organisms, their organisation and their dialectical relations with the environment has been under development by contemporary biologists. Often these perspectives have been developed in direct opposition to the kind of reductionism offered by, for example, sociobiology. Instead of over-emphasising genes in affecting an organisations development and behaviour, the emphasis is on the *potentialities* of organisms with, again, these being realised in different ways (or perhaps not even being realised at all) according to the features of their environment. Genes, according to this view, are important, but rather as carrying-codes between generations of these potentialities. In Britain C.H.Waddington is nowadays seen as the earliest biologist who was well aware of the latest developments in genetics but still insisted on understanding organisms as a whole. Again note his insistence on the interaction between an organism's potentials (in this case the potentials of a human organism) and her or his environment. By 1961 Waddington was able to develop a holistic understanding of organisms and their environment which takes into account recent developments in genetics. The first step in the understanding of heredity is to realise that what a pair of parents donate to their offspring is a set of potentialities, not a set of already formed characteristics.... Even if your parents were both Anglo-Saxons ;you do not inherit their white skin; you inherit potentialities of such a kind that if you grow up with very weak sunlight your skin will be very fair in colour, while if you are frequently exposed to much stronger sun it will be considerably browner. Nowadays we use the word 'genotype' for the collection of potentialities which are inherited. Contrast this with the 'phenotype', which is the name for the collection of characteristics which an individual actually develops under the particular circumstances in which he happens to grow up. Any one genotype may give rise to many somewhat different phenotypes, corresponding to the different environments in which development occurs (1961, p. 29). Organisms, seen in their dialectical relationship with the environment is now a central theme in a small, but growing, branch of biology. As Ingold (1990) puts it, 'organic forms come into being and are maintained because of a perpetual interchange with their environments, not in spite of it'. As will be clear from this, the argument represents something of a departure from the more classical view of evolutionary biology, whereby the organism is seen as necessarily adapting to a pre-given environment if it is to survive and thrive. 'The environment' (which after all is composed of other organisms) is both formed and formed *by* the organism. Thus Wesson (1993) argues that now 'one might better speak not of the origins of species but of the development of ecosystems' (p. 157). What is taking place during these dialectical interactions between organisms? As the laws of thermodynamics and ecological theory tell us, it is of course the transfer of energy between organisms. Energy, its creation, transfer, loss, storing and (following the death of an organism) its recycling back into the flourishing of future organisms is the key way of envisaging the *connections* between organisms. Ecosystems are of course the means by which these processes take place (see, for example, Lee, 1988) with organisms at different trophic levels operating as nutrition and energy for those at a higher level. And, as is well-known, some organisms do not fit conveniently into single trophic level. A grizzly bear eats and grows as a result of feeding of lower-level organisms such as roots and berries. At the same time, it consumes small animals and fish at higher levels. Meanwhile, all organisms and their waste decompose and turn into nutrients for the lowest level of the trophic system and all organisms produce heat. The environmental debate is largely, of course, about how one particular type of organism has managed to feed on several trophic levels and at the same time produce levels of heat which are threatening humans and other species. The two-way, dialectical, relation between organisms is therefore the key methodological stance here. On the one hand (as for example Ho and Saunders (1986) argue) the environment impacts on to organisms. Furthermore there is at least the possibility of such impacts become hereditarily fixed or assimilated by organisms themselves. This relation does not form part of classical evolutionary theory but there is no logical reason why cells could not react genetically to signals from the environment. The processes involved may not be well understood, but everyday experience suggests that this kind of interaction between organism and environment is taking place. Plants, it seems are prone to act flexibly according to environmental conditions. Aphids for example are what Wesson calls 'very plastic' in their genetic expression. Depending on season, abundance of food, kind of plant on which they live, and even the attention of ants, they may reproduce sexually or asexually, lay eggs or give birth to large mobile young, be winged or wingless and vary in size and shape. A single species may have as many as twenty different forms (p. 229). On the other hand, organisms (including in particular, of course, the human organism) is at the same time actively making and re-making its environment. Again, some of this is not a familiar theme in conventional biological theory; Darwinism basically arguing that organisms have to adapt to, and find a niche within a pre-given context. But as Lewontin (1982) argues, We cannot regard evolution as the 'solution' by species of some predetermined environmental 'problems' because it is the life activities of the species themselves that determine both the problems and the solutions simultaneously. Organisms within their individual lifetimes and in the course of their evolution as a species do not *adapt* to their environments, they *construct* them. They are not simply objects of the laws of nature, altering themselves to the inevitable, but active subjects transforming nature according to its laws (p. 160). Interestingly, however, Wesson suggests that some organisms may be more robust than others in adapting to environmental change, some of which they have created themselves. It seems likely that more complex species (especially those with highly developed nervous systems which allow them to plan and adapt more effectively) are more likely to be flexibly self-governing. Arguably this is where humans have 'triumphed' in their relations with the rest of nature. As a species they have been best able to adapt to environmental change; largely throught their capacities for developing, storing, communicating and applying knowledge. This has led to their long-term survival. And, at least until recently, they have used and concepts and ideas which largely focus on the survival of humans (as distinct from other species) and assume that humans will remain immune from the environmental change they are carrying out.

So this chapter is suggesting that key features of this approach to biology and related sciences (specifically that of organisms both changing nature and 'unfolding' their potential in relation to nature) has clear philosophical overlaps with social science, one which sees *humans* as both changing nature and developing in relation to the nature they have modified. This is a dialectical approach to understanding. It commends itself by beginning to over-ride the rigid division of labour between the natural and the social sciences. It leaves open to empirical analysis the understanding of precisely *how* this dialectical relation works out in practice. It is also a conceptualisation which, I would guess, lends itself to the kinds of 'practical knowledge' outlined earlier in this chapter. The notion of all organisms as 'unfolding', or flourishing (O'Neill, 1993) and eventually being recycled into future life is one which could recommend itself to non-experts and to their day-to-day understandings of the non-human *and* human worlds. At the same time it would link with, and breathe life into, a relatively lost tradition of more abstract concepts in the the natural and social sciences.

Developments in the natural sciences confirm Marx's early idea that working on nature changes the concepts used to understand that same nature. As we have seen, the emphasis is now turning towards the development of whole organisms and away from reductionist concentration on genes. Against this, however, there are features of these new approaches in biology which are not so helpful. There is a tendency within some of this work to over-reify wholeness and its relation to parts. Indeed, this is a recurrent feature of much 'green' philosophy, one which places enormous stress of ecological and physical systems as a whole and their impact on individual parts such as organisms. Thus Sheldrake (1990), for example, argues that humans and other organisms are subject to invisible fields. And it is a short step from this to argue that all organisms are influenced by ley-lines, or systems of energy which link up all living beings with the rest of the Cosmos. At this point the dialectical relation between organisms' potentials and their environment starts turning into something over-mystical. Individual organism are becoming 'structured' by external events and processes. The analysis is reminiscent of the 'structuralism' that infected Marxism in the 1960s and 70s. It also calls to mind Marx's critique of religion; 'holism' being a human-created phenomenon which then starts to dominate the actions and thoughts of the very individuals who created it. It also, of course, draws attention away from such social relations as class and gender. Remaining wary of these difficulties, the proposal of this chapter is to rescue those elements of modern biology concentrating on whole organisms and to deliberately neglect others. This done, whether it is humans (composed of certain biological-inherited needs and potentials) changing, and by being changed by their 'environment' or whether it is essentially the same dialectical process occurring to any organism, we have in this approach to biology the beginnings of a science which links all organisms to one another. This link is in primarily *philosophical* terms; insofar as it is a general understanding of how organisms are ordered, how they *self*-govern and how they relate to one another (see also Kauffman, 1993). As such it begins to undermine the society-nature dualism which has plagued much environmental thinking as well as enable us humans as undergoing processes which are essentially similar to those of other species. This type of understanding is perhaps the main inheritance to take from Marx. The next stage is to explore the relations between organisms and environment in the modern social-cum-environmental world. Do such relations become, in Marx's terms, simply alienated? Are there indeed ways in which humans can re-gain their lost relation with the non-human world?

4. HUMAN SOCIETY AND NATURE: THE DIVISION OF LABOUR AS A KEY OBSTACLE

This chapter has so far argued for an alternative understanding or organism-environment relations, one in which organisms and environments are dialectically related and interdependent. We now come to the question of the relation between human society and nature in modern human society. The basic issue here is whether our human society enables people to gain an understanding of the dialectical relations between people and nature to be developed. Or does it restrict such a relationship? By 'modern society' I refer to the production and consumption of commodities on the one hand and, especially in the previously state-socialist societies, the direction of economies and management of resources by bureaucratic organisations. Is it, then, the case that modern society necessarily alienates people from the natural world, or should it be seen as underpinning and encouraging this relationship? The answer is both! The market (specifically, the selling and purchasing of goods) can be envisaged as what Hayek (1988) calls the 'extended order.' It is a way in which humans can continue to some kind of changing relationship not only with other people but with the natural world. And, as enthusiasts for the attaching of market values to the environment insist, the market can be seen as a way in which societies continue adapting to the changed environmental conditions which they themselves are creating (Whelan, 1989; Lal, 1990). It is not necessary to outline the specifics of this connection in detail. Market values are attached to the environment. Pollution of the environment, the use of resources and the purchase of various forms of 'green' commodity (including 'eco-tourism') are familiar cases in point. The market can, therefore, be seen as the predominant modern way in which humans and their needs for a relationship with the environment can be seemingly recovered and retained. In combination with other relations with nature in the sphere of 'civil society' (defined here as the sphere of social life outside capitalist production and the state) it facilitates some unfolding of what Marx called 'man's natural and species being.' It allows people (and some people more than others) to 'unfold' in relation to the physical and natural environment. And, as proponents of the market since Adam Smith have argued, it operates as a coordinating device expressing individual wishes to the producers of commodities. Again, the market can be combined with other institutions such as governments and all kinds of voluntary association in forging and retaining a direct link between humans and nature, what Marx called their 'inorganic body'. This necessarily entails asking people how much they would be prepared to pay for some aspect of the environment to be preserved or improved. This means economic values can be attached to the environment and, as individuals or organisations, people can be 'charged' for their maintenance and use. Such strategies are of course widely promoted by some branches of environmental economics (see for example Pearce et al., 1989 and for practical examples Barde and Pearce, 1991). But this is surely not the end of the story. First there is the type of fetishism encouraged by the market to which Marx referred and which was briefly outlined at the beginning of this chapter. Thus humans through the market (and, more broadly, within civil society) may certainly gain some sense of re-connection with nature as they become charged for being re-attached to their 'inorganic body' or as they engage in the vast range of practices (for example gardening or going for a walk) outside the sphere of production. On the other hand, this apparent connection is of a highly fetishised form. It remains largely innocent of the material ways (particularly the use of energy and the eventual return of what is misleadingly called 'waste' into the atmosphere and soil) in which society combines in a material fashion with nature. In other words, the labour-process and the social relations of production play no significant part in this re-engagement with nature. Nature remains something external, 'out there'. It is not something made by modern capitalist and state-socialist societies which they are now obliged to live in. Rather, it is something which we can either choose or not to choose to purchase. Attaching monetary values to the environment finishes up, to use Collier's words, 'falsifying' and 'trivialising' what is being valued (Collier, 1994:5). A forest is something to be valued in its own right, rather than something which is valued 'in the heads' of paying customers who are walking through it. There is also, the secondary and obvious question as to who can and who cannot engage in this extended order connecting humans with the environment. 'Green' holidays and the extensive purchase of environmentally-friendly goods are hardly a realistic prospect for everyone, even if these *were* satisfactory ways of relating to the environment.

Thus the central point which Hayek and the market-enthusiasts miss is that human societies have to work on nature to produce commodities and they are compelled to dwell in the environment which they have created. So a view of the relationship between humans and nature in modern society which focusses entirely on attaching market values to the environment, exchanging it for money (in other words treating it like all other commodities) may be a comparatively easy way of proceeding. But such an approach, leaving aside capitalist production and the division of labour is extremely limited, not to say dangerous. It is precisely in the sphere of production and the division of labour that the dialectical connections within organisms and between organisms and their environment become systematically broken up, disjointed and misunderstood. Valuing the environment as part of the exchange of money for commodities may be a valuable first step but it does relatively little to restore these connections. As the global environment becomes increasingly incorporated into the global market the former becomes increasingly represented as the latter. But a commodified nature provides at best a limited understanding of peoples' relations with their environment.

5. CONCLUSIONS

It is in the spheres of social relations and the production of commodities that we can begin to gain a more complete understanding of peoples' (alienated) relations with nature. In his early work Marx suggested that it is the division of labour and the institution of private property that inhibit this relation. In his later work it is the labour-process and the use of nature as a mere input into production where he saw such alienation starting. Both perspectives, this chapter is arguing, are obvious starting points for a social theory to develop a connection with the environment. There are a number of different ways in which human societies work on nature (Benton, 1991b). They can, as Marx originally implied, simply use raw materials to produce some object. But they also use existing materials of nature towards human ends. In this case they are not so much transforming nature to make commodities but using existing qualities of selectively appropriated materials for such human ends as making shelter or creating food. There are other relations which can be most accurately seen as manipulating the inner powers of organisms and of the physical world towards human ends. In this third case we are talking of using and manipulating the 'built-in' propensities and tendencies of organisms; the ripening of seed and fruit, the fattening of lifestock, the use of the soil's own properties and of seasonal rhythms. Again, these are not so much instances of making new commodities but using nature's own powers towards predominantly human ends. But to a large extent, this chapter is arguing, the mismanagement of the environment (including, of course, the health-effects on human beings) is a product of the coordination problems and power-plays associated with advanced industrial societies in general and not just with processes and social relations associated with capitalism. As such they would be the result of any advanced industrial production-system. As Marx started to argue in his later work, hierarchical relations of management and the supervision of collective labour are likely to be a necessary feature of any modern industrial system. Capitalism may exacerbate the situation by heightening the divisions between different types of expertise, between mental and manual labour by rapaciously exploiting the environment and by ignoring the type of practical knowledge acquired by workers most immediately involved in the production, circulation and exchange of goods. But if the argument of this chapter is right, the underlying tragedy has been that environmental crisis is the product of modernity itself. Each independent branch of knowledge has made great strides on its own, but at the expense of coordinating with ,and understanding, each other. This applies, as I have argued, not only to the separation between disciplines (especially those of the social sciences on the one hand and the natural and physical sciences on the other) but between more theoretical knowledge and the anecdotal knowledge which people are assembling during their daily lives. This situation clearly cannot be allowed to continue and this chapter is a first step towards demolishing these boundaries.

REFERENCES

Barde, J-P. and D. Pearce. (1991). *Valuing the Environment. Six Case Studies*. London: Earthscan.
Benton, T. (1991a). 'Biology and Social Science: Why the Return of the Repressed Should be Given a (Cautious) Welcome'. *Sociology* 25.1: 1-29.
Benton, T. (1991b). 'The Malthusian Challenge: Ecology, Natural Limits and Human Emancipation'. In P. Osborne, *Socialism and the Limits of Liberalism*. London: Verso, p. 241-70.
Benton, T. (1992). *Natural Relations. Ecology, Animal Rights and Social Justice*. London: Verso.
Benton, T. (1994). 'Biology and Social Theory in the Environmental Debate'. In: M. Redclift and T. Benton, *Social Theory and the Global Environment*. London: Routledge.
Collier, A. (1994). 'Value, Rationality and the Environment'. *Radical Philosophy* 66 Spring:3-9.
Dickens, P. (1992). *Society and Nature*. Hemel Hempstead. Wheatsheaf: Harvester.
Dunlap, R. (1980). 'Paradigmatic Change in Social Science. From Human Exemptionalism to an Ecological Paradigm'. *American Behavioral Scientist* 24:5-13.
Eckersley, R. (1993). *Environmentalism and Political Theory. Toward an Ecocentric Approach*. London: UCL Press.
Grundmann, R. (1991). *Marxism and Ecology*. Oxford University Press.
Haldane, J. (1913). *Mechanism, Life and Personality*. London: Murray.
Haldane, J. (1917). *Organism and Environment as Illustrated by the Physiology of Breathing*. New Haven: Yale University Press.
Haldane, J. (1935). *The Philosophy of a Biologist*. Oxford: Clarendon Press.
Hayward, T. (1994). 'The Meaning of Political Ecology'. *Radical Philosophy* 66 Spring: 11-20.
Hayward, T. (1995 forthcoming). *Ecology and Enlightenment*.
Ho, M-W. and P. Saunders. (1986). 'A New Paradigm for Evolution'. *New Scientist* 27 Feb.
Lee, K. (1988). *Social Philosophy and Ecological Scarcity*. London: Routledge.
Lewontin, R. (1982). 'Organism and Environment'. In: H. Plotkin (ed.), *Learning, Development and Culture*. Chichester: Wiley.
Lewontin, R. (1993). *The Doctrine of DNA. Biology as Ideology*. Harmondsworth: Penguin.
Ingold, T. (1990). 'An Anthropologist Looks at Biology'. *Man* 25: 208-29.
Kauffman, S. (1993). *The Origins of Order*. Oxford University Press.
Lal, D. (1990). *The Limits of International Cooperation*. London: Institute of Economic Affairs.
Manicas, P. (1987). *A History and Philosophy of the Social Sciences*. Oxford: Blackwell.
O'Connor, J. (1988). 'Capitalism, Nature, Socialism. A Theoretical Introduc-

tion'. *Capitalism, Nature, Socialism* 1 Fall: 11-38.
O'Connor, M. (ed.). (1995). *Is Capitalism Sustainable? Political Economy and the Politics of Ecology*. New York: Guilford.
O'Neill, J. (1993). *Ecology, Policy and Politics*, London: Routledge.
O'Neill, J. (1994). 'Humanism and Nature'. *Radical Philosophy* Spring: 21-29.
Outhwaite, W. (1979). 'Social Thought and Social Science'. In: P. Burke (ed.), *The New Cambridge Modern History*. Cambridge University Press, Vol.XIII p. 271-92.
Pearce, D. et al. (1989). *Blueprint for a Green Economy*. London: Earthscan.
Rattansi, A. (1982). *Marx and the Division of Labour*. London: Macmillan.
Ravaioli, C. (1993). 'On the Second Contradiction of Capitalism'. *Capitalism, Nature, Socialism* 4(3): 98-104.
Sayer, A. Walker. (1992). *The New Social Economy*. Oxford: Blackwell.
Sheldrake, R. (1990). *The Rebirth of Nature*. London: Century.
Szerszynski, B. (1995). 'On Knowing What to Do: Environmentalism and the Modern Problematic'. In: S. Lash et al (eds.), *Risk, Environment and Modernity: Towards a New Ecology*. London: Sage.
Tamas, P. (1992). 'Neoclassical Development Models and Environmental Regulation: New Dilemmas in Eastern Europe'. Paper presented at the symposium on current developments in environmental sociology. SISWO, 1018 Amsterdam, The Netherlands.
Waddington, C. (1961). *The Nature of Life*. London: Allen & Unwin.
Wainwright, H. (1994). *Arguments for a New Left*. Oxford: Blackwell.
Wesson, R. (1993). *Beyond Natural Selection*. Cambridge: MIT Press.
Whelan, R. (1989). *Mounting Greenery*. London: Institute of Economic Affairs.

NOTES

1. O'Connor's arguments on the contradiction between capitalism and nature have been extensively discussed in subsequent issues of *Capitalism, Nature, Socialism.*

3. Fragmentariness and Interdisciplinarity in Linking Ecology and the Social Sciences

Andrej Kirn

Abstract

Modern interactions between nature and society and the current types of environmental problems offer new possibilities and incentives for the linkage of natural and social sciences. Ecology as a natural science is less and less able to observe processes in the biosphere independently of human activities. Environmental problems not only present an opportunity for expanding the sphere of sociological and social research in general, but also represent a theoretical/paradigmatic challenge for the social sciences.

There exists not only an intellectual and epistemological need, but also a praxeological need for certain levels of unity and holism in science. Fragmentary and disciplinary science also implies fragmentary and disciplinary practice. Ecology has explicitly shown the holistic character of the biosphere and ecosystems, which requires a holistic and interdisciplinary way of thinking. The theoretical formulation and practical realization of the concept of an ecologically sustainable society and ecologically sustainable development will demand and encourage not only multidisciplinarity, but also interdisciplinarity in science, especially as regards the social, natural and technical sciences.

1. INTRODUCTION

The relationship between social and natural sciences is the most critical part of the entire discussion on the unity of science because of the ontological gap between them. Three fundamental answers are available:

1. The naturalistic/positivistic tradition stresses the identical logical structure of explanation in both natural and social sciences. For Nagel (1961), not a single methodological difficulty is characteristic only of the social sciences, though some are quite serious. Others, again, reject the epistemological distinction between natural and social sciences using the criterion of exactness, as if to say that both fields are inexact in certain dimensions and levels. The naturalistic position views the existing special status of social sciences as a transitional stage in its development from soft to hard science. Giddens (1976:13) dismissed such utopian expectations with the following witty metaphor: 'But those who still wait for a Newton are not only waiting for a train that won't arrive, they are in the wrong station altogether.'

On the contrary, Turner (1991:585) is very optimistic, saying 'that sociology can in principle be a natural science'. Here it is very important to know which historical type of natural science is meant. If one has in mind the classical Descartes/Newton paradigm of natural science, then one can fully agree with Giddens' metaphor. But if one refers to the emerging paradigm of postclassical or postmodern natural science, then Turner's expectations are indeed not unfounded. The postmodern paradigm, whose beginnings go back to the second half of the 19th century, introduced a new understanding of determinism, of the relationship between the subject and the object of research, part/whole, chaos/order, stability/instability, evolution, time, reversibility/ irreversibility, etc. The introduction of new categories and the conceptual modifications of old, fundamental ones offers new opportunities for the integration of natural and social sciences. Postmodern science is no longer merely a platonic science of being, but also a Heraclitean science of becoming and passing away. Not only natural science, but social science as well is entering into a new dialogue with nature. And human beings are simultaneously entering into new, practical relations with nature, relations which represent new challenges and involve new risks. Postmodern natural science constructs images and models of nature that not only stir our sensual perception, but also challenges the abilities of our imagination. Postmodern natural science constructs an astonishing, paradoxal world which it is increasingly more difficult to comprehend and describe. This is not the end of rationalism, but merely a transition to a new and more demanding form of rationalism, a new and more demanding way of thinking. Postmodern science abandons and goes beyond many traditional dualisms. Something similar is happening with a regard to some sociological theories, for example in Giddens' (1987:220) concept of the 'duality of structure'. If sociology still had the paradigm of classical science as a model, it might very well have had the same fate as economics. In its theoretical models, the neoclassical political economy took advantage of certain principles of and analogies with classical mechanics of the 19th century (law on the conservation of matter and energy, equilibrium, reversibility, energy). However, after physics introduced significant theoretical innovations at the turn of the century, economics discontinued its productive dialogue with natural science (Mirowski, 1989; Roegen, 1971).
2. The antinaturalistic tradition stresses the major differences between the two fields. Methodological/theoretical imitation of natural science by social science is unproductive and wrong. The rejection of naturalism is not reason enough for the rejection of any linkage of social sciences with natural sciences.
3. In my opinion, a productive and promising approach is one that does not advocate the identity of nor total discrepancy between natural and social sciences, but searches for general differences and similarities, or even identities, in different dimensions: subject/object of research, character of a law or regularity and specific aspects of an experiment, heterogeneity/homogeneity of an object in time and space, measurability of phenomena, absence of numerical constants in social science, the role of values and goals, etc. (Machlup, 1969). A cardinal ontological difference exists in self-reference, self-observation, self-description

of a social subject and social system, as well as in the appropriateness of human behaviour. Individuality, for example, is not only present in the social universe, but in nature as well. The biological world is very heterogeneous, yet it does not possess such time dynamics of heterogeneity as social development does. If typological repeatability is disregarded, then concrete social events are unique and unrepeatable. Even certain physical events or processes are, at least from the aspect of today's knowledge, unrepeatable. These are, for example: the big bang, the creation of Earth, the formation of continents, etc.

Within the framework of contemporary interactions between nature and society, the linkage of social and natural sciences could be more realistic and epistemologically more successful today than it was in the past[1]. Numerous homologies, analogies and metaphors transferred from biology to sociology (organism, division of work, struggle for survival) did not prove justified and are more or less of historical interest today[2]. Sociology was linked to biology from the very beginning. Comte even expected that 'Sociology will in the future... (provide) the ultimate systematization of biology.' (Turner, 1991:152). Recent endeavours have taken the opposite direction of Comte's expectations. According to Wilson, sociobiology is 'the systematic study of the biological basis of all social behaviour' (Turner, 1991:152). Many types of social behaviour have very little or nothing to do with biological processes. On the other hand, sociologists and biologists shouldn't forget that much of what appears as a pure biological fact has already been socially/historically conditioned and transformed. The social is realized, embodied in the biological. Our bodies and minds are in essence a social/biological fact. Human history is embodied and transformed in the evolution of our bodies and our minds. And because the social exists as the biological, this further complicates the relationship between human biology and sociology, psychology. Despite the present difficulties and hesitations accompanying the linkage of genetics and cultural evolution, I am deeply convinced that further efforts in this direction can lead to persuasive and surprising results. 'Gene-culture coevolution' is just one aspect of the complexity of the 'nature-society coevolution'. In principle, one should allow for the possibility that there can be exceptional epistemologically productive connections and convergencies between various disciplines.

It is often unjustly believed that only social science has benefited from the methods, analogies and metaphors taken from natural science. History shows that precisely the reverse process was strong and usually more productive: for example, the influence of Quetelet's social statistics on statistical physics, the influence of the notion 'division of work', which originated in the political economy, on the concept of the physiological and ecological division of work in biology. Mandelbrot's idea of fractals grew out of economic research. The successful application of homologies, analogies and metaphors occurs when these are adapted in full consideration of the specificities of the new context. Yet the function of concept transfers may vary considerably. 'They may serve as heuristics and theoretical construction tools, as pedagogical and persuasive devices, as polemical weapons, as legitimizing label or as evidential support.' (Limoges, 1994:336). Marx, for example, used analogy in the setting of a scientific task: it would be necessary to write the history of the productive organs of the social human being just as Darwin wrote the history of the development of the productive organs of plants and animals. In the 19th century the analogous, homologous and metaphoric use of the terms of natural science in sociology and economics was also stimulated by the endeavours of these sciences to constitute themselves as a science according to the standards formulated by natural sciences, in particular physics. The success of a particular science and its extensive intellectual and cultural influence was reason enough to stimulate the testing of its concepts and methods elsewhere as well.

2. ENVIRONMENTAL PROBLEMS AND THEORETICAL/PARADIGMATIC CHALLENGES

The development of natural sciences, specially ecology, the theory of self-organizing autopoietic systems and practical environmental problems present different possibilities for the linkage of social and natural sciences in the 20th century than there existed in the 19th century. Today one speaks of the invasion of ecological dimensions into various sciences. This inclusion of ecological issues has two basic phases.
1. The formation of new research fields, themes or even specializations (e.g. environmental law, environmental ethics, social ecology or environmental sociology, environmental management, environmental economics, etc.) within the scope of existing disciplines, yet without a pronounced influence on the modification of their fundamental ideas.
2. The modification of theoretical bases and paradigms of the paternal disciplines.

2.1 A challenge for ethics

When environmental ethics began to take shape in the 70's (if we ignore its numerous predecessors), two fundamental answers to the question of the relationship between the so-called anthropocentric and ecocentric (biocentric) ethics were formulated. According to one version, environmental ethics is merely an extension of the existing ethics to include plants, animals, ecosystems, i.e. nature as a whole. This expansion could not represent a theoretical challenge to ethics, just as various professional ethics do not. A more radical interpretation says that this not only involves widening the field of ethical judgment, but modifying the ethical paradigm as well. The paradigm of anthropocentric ethics was based on the key axiom that morality is the relationship between free and intelligent subjects. Environmental ethics expands the field of moral objects to include those that are neither free nor intelligent as human moral subjects are. This situation radically changes the relationship between a moral subject and object. One is faced with the question why, indeed, should a human being have moral considerations and obligations towards natural entities? Once again, the problem of the relationship between intrinsic and instrumental values comes to the fore. In reality, environmental ethics demands a new reflection of man's attitude towards all forms of life and stresses the ultimate limits of ethical responsibility for the ecological conditions of future generations. Some believe that, in the long run, it makes no difference whether certain natural entities are protected because of the intrinsic values attributed to them by humans or on the basis of their instrumental value for man. From the physical, ecological aspect it really doesn't make any difference, but it does make a difference from the aspect of man's spiritual attitude and values. I believe that environmental ethics represents a challenge for the existing ethical tradition.

2.2 A challenge for the economy

Similar questionable relationships exist between the political economy and ecological or environmental economics. Environmental economics has assumed the relevant findings of ecological as well as other natural sciences, in particular physics. It has recognized the external costs caused by pollution and has directed attention to the utilization of natural resources. It is difficult to say for certain whether environmental economics represents an antithesis to neoclassical economics or if it is more or less a new phase in its continuous development (Hampicke, 1992:303). If environmental economics is not understood solely as the technical application of the notions and methodological apparatus of economics in dealing with environmental problems, cost-benefit analysis and the economical use of natural resources, but rather as a much wider understanding which include growth and development, the intergenerational utilization of natural resources and the use of the entropy metaphor, then we can with good reasons also speak of the tendency towards constructing a new paradigm. Georgescu-Rogen (1971) formulated the basic outlines of the entropy paradigm of an economic process which, in terms of deep and shallow ecology, one could say represents deep economy. For an economy that only reflects environmental problems through external costs and economical use of natural resources, this primarily involves important pragmatic corrections. Daly and Cobb (1989) gave an opportunistic epistemological reason for the inclusion of external environmental costs. 'Externalities do represent a recognition of neglected aspects of concrete experience, but in such a way as to minimize restructuring of the basic theory.' (Daly and Cobb, 1989:30). The dominating abstractions become an obstacle in observing and recognizing that which has been neglected. Even the method, not only the theory, selects an object and influences what it perceives and recognizes as significant. The social institutionalization of disciplines additionally strengthens the barriers. In a world of predominant discipinarity, the distrust in synthesizing the construction of reality is quite understandable. Disciplinarity isolates individual aspects of totality and develops methods which further reproduce and intensify this isolation[3]. Theoretical abstraction is often supported by the real abstract. Not only theory, but real economic practice did not consider external costs for a long time. The real abstract was a basis for the notional abstract. Theoretical recognition of external costs preceded the overall, yet limited, practical recognition. Once the externalities grew to such an extent that their strict consideration could not only mean profitless operation, but also the liquidation of entire economic branches, externality actually became an internal defectiveness of the economic system itself. It is necessary to introduce new abstractions which also encompass what previously had the status of the external. Externalities are a temporary aid used to evade the defectiveness of a theory that ignores the connections of an economic system with nature. The fact that the economic system is a subsystem of the 'natural economy' has still not received full theoretical recognition. The human economy can only be maintained within this broader system, despite the scientific and technological potentials available for transforming nature. Locke's marginalization of nature in social and economic theory was the harbinger of a theoretical and practical process which dominated the past 300 years in the period of industrialism and the modern era. With the famous assertion that 99/100 of the value of manmade products can be attributed to work, Locke (1964:314) announced that we were entering an epoch in which nature was subordinated to the economy and human labour. The value of nature approaches zero, while the value of human labour grows infinitely. Locke's degradation of nature is still practised today, when the value of natural resources in the GNP amounts to only a few percent. The significance of nature has been narrowed down to resources and to the general conditions of human life. The use value of nature and its free services is only as high as it can be expressed in monetary form. Although in his labour theory of value Marx attributed the source of value only to work, he attributed the source of material wealth (use values) to both work and nature. Neglecting nature's contribution to material wealth was even branded by Marx as the 'divinization of human labour'. From the ecological aspect, the economy has still not managed to successfully join both descriptions, the physical and the monetary. 'In contrast, most econometric models contain no representation of physical process, but only human production and consumption behaviour defined in economic (i.e. example) monetary terms.' (Robinson, 1991:636) The goal of considering physical dimensions is not to change economics into a technology or natural science. Physical/energy values will only complement and not replace monetary values. It will never be possible to express the value of most of nature's services in monetary units, yet simply recognizing and considering the economic value of various natural resources represents a giant step towards the economically more premeditated utilization of these resources in comparison with the situation in which their economic value was not recognized. A new status of nature in economic theory and the full recognition of natural bases and conditions for a sustainable human economy are necessary.

2.3 A challenge for sociology

For sociology, environmental problems not only represent an opportunity for expanding the subject of research, but also a theoretical challenge. Sociological research has not limited itself to dealing with the human perception of and reaction to environmental problems and man's value orientations, but has attempted to penetrate into the social/ecological content and implications of the physical activities of humans and their needs. And here sociology stumbles upon the same or even greater difficulties than economics when it comes to linking the physical and nonphysical content of human activities. This linkage is often only possible by means of mentally reconstructing and combining various information. Finding out how, for example, environmental problems represent a threat to democratic tradition or national wealth is a task which cannot be solved by any empirical study, nor by a predominantly empirical one, which, on the other hand, does not mean we are ignoring empirical findings. The connections between the forms of appropriation of nature on the one side and such social phenomena as, for example, the social division of labour, social structure, international trade and communications on the other, can only be reconstructed and constructed using the mental/theoretical method. Some sociologists are contemplating the global political implications and consequences of an eventual ecologically caused scarcity society for the existing modern democratic tradition and welfare state. Or, let us consider the problem of human freedom, which certainly does not have only political and economic dimensions, but ecological dimensions as well. How does man's relation to nature constitute freedom? No empirical study can provide a synthetic answer to this question. On the other hand, various empirical studies can show how the space of man's freedom narrows in a concrete historical situation because of ecological conditions. According to Sorokin (1957: 487), man's *'freedom becomes restricted when he cannot do what he would like to do; has to do what he would prefer not to do; and is obliged to tolerate what he would like not to endure.'* With the rise of industrialism, the conviction that man was almost completely independent of nature became generally accepted. This conviction found its support in:
• the growing scientific/technical achievements that opened the way to new natural resources, successfully realized many of man's goals and set new, more ambitious ones;
• the growing social division of labour, causing the largest share of the active population working in the secondary and tertiary sectors to lose contact with nature;
• the change of direct dependence on natural resources in satisfying numerous needs into a socially conditioned indirect dependence. For the majority of people, dependence on nature is realized through work and the products of others. Direct dependence on nature was inevitably replaced with an increasingly more intensive and extensive dependence on people;
• the attempts of science and technology to make industrial, transport and even agrarian activities as independent of the changing natural and climatic conditions

as possible;
- the less-expressed dependence of industrial activities on and satisfaction of human needs from local natural resources as the consequence of international trade. The transparent local dependence on natural resources changed into a nontransparent, global planetary dependence. For consumers, international trade wiped out the connection between ecological consequences and consumer products. Who among those millions who ever bought a hamburger is even aware of the possible connection between hamburger meat and the transformation of forests in Costa Rica into pastures, which may in some years turn into a wasteland?

The specific epistemological requirements arising from the linking of the social and physical/ecological dimension could have been the reason for narrowing the horizon of possible social/ecological analyses, or for theoretical deficits in this field. Such narrowing, of course, cannot lead to the formation of a 'new ecological paradigm' (NEP). As formulated by Catton and Dunlap (1980), NEP essentially surpasses the sociological frames and has a broader social, cultural and civilisational meaning. Catton's and Dunlap's NEP do not deny the principles of the 'human exemptionalist paradigm' (HEP), but include them as complementary and subordinated elements. They admit, for example, that: 'While humans have exceptional characteristics (culture, technology, etc.), but they remain one among other many species... Human affairs are influenced not only by social and cultural factors, but also by intricate linkages of cause, effect and feedback in the web of nature...' (Catton and Dunlap, 1980:34). What I miss the most among the listed properties of NEP is the entropic characteristic. NEP should recognize the real significance of the entropy law in social material/energy processes. How and under what conditions the entropy law can be applied in the sociological field and how to adequately formulate the term 'social entropy' and its limits are the questions which sociologists, contrary to economists, have yet to tackle. The very existence of environmental sociology denies Durkheim's axiom that social facts can only be explained on the basis of social facts. The question is, of course, how broad our understanding of a social fact is. Are cultivated fields, roads, automobiles, etc. merely physical facts or are they social physical facts as well? They certainly aren't natural facts, nor are they social facts in the same way as religion, moral norms, social institutions, social groups, symbolic interactions and communications. Even the hole in the ozone layer is no longer a natural fact. All the anthropogenetic ecological consequences and all the transformations of nature are not simply natural physical facts, but social natural facts as well, a linkage of the social and the natural. For environmental sociology, it is very important how the notions environment and nature are understood. Environmental sociology must limit the notion environment to the physical environment in order to expose that dimension of the environment which usually is not considered in the cultural, political, social and economic environment.

Neither Spencer's nor Durkheim's sociological traditions represent favourable theoretical possibilities and incentives for the transition from the 'human exemptionalist paradigm' to the 'new ecological paradigm'. Peter Dickens believes that Marx and Engels are the only writers who developed the science presently needed for understanding environmental issues. '... their ideas would be the best basis for the development of Catton and Dunlap's "new ecological paradigm".' (Dickens, 1992:XIV) I think that Marx's social and anthropological theory also contains elements making Marx's standpoint ambivalent and questionable from the ecological aspect. Marx adopted the unrestrained development of productive forces and only rejected its capital social form. Once it loses its capital social form, the universal development of productive forces is, for Marx, nothing more than the universal appropriation of nature, the universal development of man himself, his abilities, pleasures and needs. Marx accepts the civilizational role of capital in relation to nature, yet at the same time he warns that capital destroys both sources of wealth: the land and the worker. In comparison with the 19th century, the destructive role of capital in developed, capitalist countries has softened with respect to the material, social and educational status of workers. Environmental legislation and standards have changed or even reduced the degradative effects of capital on nature, however, on a planetary level the destructive load imposed by capital on nature has not been lowered, but even increased. Marx and Engels actually did not know of or recognize any external biophysical environmental restrictions in social development, only social endogenous capital limitations. The entropic nature and entropic implications of social labour remained unknown to Marx. In an environmental sense, Marx views the work process as an exchange of substances between nature and society in the sense of the first law of conservation matter-energy and not in the sense of the second law, the law of entropy, which constantly generates entropic degradations representing undesired and very frequently also unexpected and unintentional consequences. Engels was closer to this idea when he discovered that, in the long run, humans always have to deal with unpleasant consequences in their victories over nature. Marx's anthropology is the glorification of human creativity and man's universal appropriation of nature. In fact, these elements are in dissonance with the ecological paradigm and point to ecological ambivalence and the deficiency of Marx and Engels' social theory. For this reason Dickens' addition to his general assessment of Marx's and Engels' ecological potential, in which he says that 'their ideas can now be simultaneously criticized and developed' (Dickens, 1992:XIV) seems quite appropriate. In order to fill the new ecological paradigm with the richest and most diversified yet coherent content, it will be necessary to attain advancement in both the theoretical and empirical disciplinary and interdisciplinary research of various environmental problems and consequences, both local and global, actual and possible.

3. JUST LINKAGE, OR THEORETICAL INTEGRATION AND UNIFICATION AS WELL?

Fragmentary and disciplinary knowledge implies fragmentary practice. The human need for various levels of holism is not only epistemological and psychological, but also praxeological. Both society and nature are a seamless whole. The more or less rough seams in this whole are made by man's disciplinary and super-specialized structuring of knowledge. If the diagnosis for environmental problems is multidisciplinary (economic, technological, demographic, axiological, etc.), then the therapy must also be multidisciplinary. Yet technological monism both in the diagnosis and in the therapy is quite frequent. If we ask ourselves why the social-economical and value system prefers certain technologies that cause specific environmental problems, we may conclude that the direct technological reasons are socially generated and selected. As regards the factors responsible for generating environment problems, it is impossible to empirically prove the direct empirical connections between them. It is only possible to give a more or less convincing explanation. The contribution of anthropocentric and theocentric ethics in the formation of a modern, practical attitude towards nature will never be precisely localized and determined. Theoretical social/ecological analyses and reflections frequently cope with such limitations. However, whatever cannot be quantified is not necessarily insignificant.

The environmental problems of modern civilization are social/anthropological by their origin. Man appears as their generator and as their victim in the final instance. Anthropogenetic environmental problems have three basic dimensions: social, natural and technical. Why does man transform nature? What kind of interests, needs, motives, troubles, ambitions and values drive him to do this? What are the social/organizational forms of changing and appropriating nature? All this is the subject of social science. What are the biophysical, medical, chemical, geological, pedological and other changes - this is the subject of natural science. What are the consequences of these changes for society, its responses to them, and what kind of strategy should be employed in solving and preventing them - this is the task of social and technical sciences, but ultimately also of the natural sciences, inasmuch as their findings represent the broadest frame for outlining technical solutions. Ecology as a natural science has laid great emphasis on the holistic way of thinking, which has an extrapolative value for other disciplines as well. In its field it affirmed and confirmed the old philosophical truth that everything is connected with everything, and this is all the more true in the case of a whole such as the biosphere. In the biosphere it is impossible to separate biotic from abiotic elements, the local from global processes, the system structural and functional characteristics of the biosphere from local and regional ecosystems. Ecology has corrected the previously one-sided picture of life's adaptation to its environment. Life also creates its own conditions and adapts itself to them. The present chemical structure of the atmosphere is to a great extent the product of the vital activities of organisms. Ecology, originally defined as the science of the interaction of organisms with their environment, also implicitly dealt with the interaction of the human community with the environment. From the biological aspect people are organisms, too, but man has created other ways and means of interacting with the environment (conscience, norms, technology, farms, factories).

People are linked to nature directly in a physical/biological sense and indirectly in a physical sense through production and consumption. At the same time, however, they are also connected spiritually through their conscience, language, communications. According to Robinson (1991:634), 'an integrated view of the relationship between human and natural systems encompasses both physical and actors views.' 'Actor view' includes relations between individuals and groups, and relates to the information, communication and decision-making processes dealing with environmental problems. The main difficulty is not the fact that human activities (demographic processes, production and consumption, resource extraction, cultural, recreational and transport activities, etc.) are described in physical terms and that we are faced with ecological loads and the consequences of these activities, the main problem is the 'integrated view' of the physical and nonphysical side of human activities. Robinson's 'integrated view of natural and human systems' is essentially only a coordination, an interconnection of two independent descriptions and not an integration in a qualitatively and categorically more demanding sense. The goals of Robinson's integrated view are primarily pragmatic. 'Natural and human systems must be represented in the form of information to be able to be used in human decision-making.' (Robinson, 1991:632). Researchers dealing with different environmental disciplines would quickly agree on the issue of target unity 'to promote good lives within a good environment.' (Sagoff, 1990:385). According to Sagoff, it is necessary to organize 'environmental science around a conception of human beings in nature, as we organize the social science around a conception of human beings in society.' (Sagoff, 1990:386). A more complex understanding of the integrated view in an epistemological sense requires categories, principles and theories which cover ontologically different processes, structures, properties, relations and systems. If we don't have such abstract formations, then all that remains is unification on the level of goals and the linkage of the physical and actor view within the frame of interacting natural and human systems. Science produces categories, principles and theories covering various ontological spheres: system, evolution, entropy, productivity, diversity, carrying capacity, biosphere, bio-socio-technosphere, catastrophe, etc. The advancement of science and new forms of interaction between nature and society will generate new linking and integrative notions and principles. The global problems of mankind and the growing interdependence of the global and the local influence both the scientific and the lay conscience to think in more global categories. The daily fulfilment of numerous human needs depends on the global division of labour, international trade, transport, communications. In the literally physical sense, this interconnection changes the local existence of people into a global one. In future, historical epistemology will probably discover and reconstruct the connections between real processes of globalization, which have been intensively progressing since the industrial revolution, and the processes of globalization of thought and conscience.

In addition to their general meaning, the above-mentioned integral notions have a specific meaning when they appear in different ontological circles. The explanatory power of these notions usually depends on the formulation of their specific meanings. Direct use in the general sense must develop into a growing specification. The theory of autopoietic and self-organizing systems carries great potential for the abstract unification of biological systems. Further development was oriented toward the specification of social systems in relation to other biological systems[4]. Here we are faced with the following paradox: A high level of abstraction enables transdisciplinary use, or the direct homologous, analogous or merely metaphoric use of these notions. The use of a general notion in ontologically different contexts actually requires a specification, a new notion within the general notion. However, this weakens the connective, integrative power of the general notion. The problem that arises is: what remains of the common epistemological content of a general notion upon its frequent, specific use? Let us take, for example, the terms entropy, system and evolution. The notion entropy has gone beyond its initial physical/thermodynamic meaning. Today it has linked with such notions as information, life, order, structure, organization, evolution, etc., and appears in at least four contexts: thermodynamic, informational, biological and social. In the biological and social context we are speaking of entropy on a macroscopic and not microscopic molecular level. Except in economics, entropy does not frequently appear in the social context or, if it does, merely metaphorically, occasionally and not half as theoretically supported and justified as in economics. Man is an entropic, creative, cultural being who creates his own social institutions, culture and structures on account of the entropic degradation of matter and energy, and the reduction of biological diversity in nature. As regards the term social system, its special features, such as self-reference, self-observation and self-description, are more important than the general difference between system and environment, the hierarchical character of a system, system-subsystem, etc.

4. EVOLUTION AND TECHNOLOGICAL INEVITABILITY OF ENVIRONMENTAL PROBLEMS

Evolution covers very diversified phenomena. Besides the abstract notion of evolution there are also very different, specific notions of evolution. From the aspect of entropy, the evolution of the sociotechnosphere differs substantially from evolution of the biosphere without the sociotechnosphere. Because of circular interdependence, life makes constructive use of mutual degradation, death. The rise of human technology has changed everything. Since the industrial revolution, human technologies have not been as cooperatively and functionally included in the remaining network of life as the 'natural technologies' of plant and animal organisms. With exosomatic evolution man has placed himself at the top of the list of entropy growth accelerators (Roegen, 1971). The general inevitability of socially and technologically generated entropy always exists in a certain historical form. Technologically generated entropy is a natural necessity, but the speed, extent and form of its growth are not. These depend on the social nature of production and consumption, on the development of science and technology, on the social realizations of scientific and technical opportunities and, ultimately, on social perceptions of the values and quality of human life, resulting more or less serious ecological consequences and burdens for natural resources. In this relationship between general necessity and the social form of entropy man has the freedom to minimize the dissipation of energy and matter through the utilization of knowledge, a selection of adequate goals, suitable forms of behaviour and ways of life. When the extent of generating a specific form of entropy (degradation and pollution) is no longer socially acceptable, strict standards are employed to reduce its effects, or it is replaced with another form of entropy, which at least in the beginning appears socially more acceptable, less risky and less disturbing. Social resistance to entropy generation is often postponed in time by dispersing it in space. When science is supported by social pressures and the willingness of the social system to take preventive action towards generating entropy, the possibilities for partial victories are, of course, unlimited. However, there is no model for a total, final victory, i.e. that the human race could, in future, possess technological capabilities of civilization that would not generate any entropy. Within the context of constant economic growth and the increased saturation of the biosphere with technics, even partial victories in the battle against entropy will become a Sisyphean task. The reduction of entropy per inhabitant, per unit of social product or per unit of physical manufacture and consumption will be almost nullified due to the growth of production. Whatever we have gained with the reduction of pollution and degradation we will lose with growth. The inability to eliminate pollution and degradation dominates the most urgent ecological and anthropological issues concerning the relationship between technology and nature, between technology and the biosphere, between technology and biological evolution. Even if technologies were introduced into the biosphere with full consideration for ecological findings, we will forever be confronted with the incompleteness of these findings. We will always be dealing with a certain degree of fragmentariness of our findings and never with their perfect, complete holism. We will never be able to fully foresee all the possible social and ecological effects of technologies. If, however, we happened to dispose with such divine foreseeability, this still wouldn't mean we were simultaneously disposing with the ability to prevent all the foreseen yet undesired effects. It is usually not possible to optimize all the goals at the same time.

Technologies are the utilization of specialized knowledge for the purpose of accomplishing specific functions and goals in such extremely complex totalities as the society and the biosphere. Interactions within and between these totalities always represent something beyond the current discoveries of science. Scientific abstractions homogenize and unify heterogeneity. This unification of heterogeneity attains its highest level in mathematical abstractions and mathematically formulated laws. These materialized, homogenized abstractions return to their heterogeneous, original form (biosphere, society, ecosystems) through technological applications The totality of original form and the scientifically/ technologically realized, specialized abstractness, which functions as a concrete, sensual abstractness, will always be in conflict. The source of epistemological and technological/economic success of science is simultaneously the source of the ecological failure of science applied in this way. In this sense the technological generation of environmental problems is the 'fault' of science. Man fills the biosphere with fully designed technological products and functions which are not formed through evolution, that is through mutual interaction with the rest of life[5]. Biological evolution does not have predetermined norms and standards for the creation of new species, while man brings technologies into the biosphere according to previously selected criteria, specifications and standards which until only recently were primarily technical and economic, not ecological. The realization of possibilities offered by genetic technology and biotechnology has not changed the existing practice, but has only expanded and intensified it in a new, delicate system such as life. Irrespective of how the role of inter- and multidisciplinary findings in the designing of technologies will grow in future, they will always be, to a certain extent, ecologically incomplete in the highly complex totality of the biosphere. Except for human beings, the remainder of life generally does not benefit from technology, but is endangered by it. Technologies have constantly narrowed their living space and transformed it to their detriment. Man's scientific and technological intervention in nature will always be the source of various types of risk. However, understanding and managing risks not only requires the cooperation of natural, social and technical sciences, but communication between the state, science and the public as well.

5. INTERDISCIPLINARITY AND MULTIDISCIPLINARITY FOR SUSTAINABLE DEVELOPMENT

The intensive linking of natural/ecological, technical and social findings is directed towards the formation of a concept of sustainable development and its indicators (Kuik and Verbruggen, 1992; Clark and Munn, 1986; Global Environmental Change, 1991; Reconciling the Sociosphere and the Biosphere, 1989), and the study of possible global climatic changes (Chen, Boulding and Schneider, 1983). Human civilization is still caught in the rails of unsustainable development. The scientific potential for transition to sustainable development is greater that the practical willingness. Today we are aware that many of our actions are not in conformity with the norms and goals of sustainable development, yet we are far from having identified all of them. Ecological studies of the carrying capacity of the environment, social/technological changes in the exploitation of the environment and their consequences will constantly reveal new conditions for sustainable development, as well as new sources of unsustainable development. Global biospheric conditions for sustainable development are necessary for every local and regional sustainable development and vice versa - regional unsustainable development endangers the global conditions for sustainable development. There is not only the tyranny of 'small decisions' (Kahn, 1966; Odum, 1982), which will always be capable of destabilizing the conditions for sustainability; there is also the tyranny of big decisions (motorways, dams, nuclear power plants, etc.). Sustainable development is a great civilizational, integratively epistemological and practical task. And because sustainable development is not only a concept of biological survival, but of maintaining permanent conditions for cultural and human life in the biosphere, it encompasses various conflicting interests, illusions and unfounded ambitions for the free and social construction of nature with the help of science and technology, all compromisingly mixed and interconnected.

6. CONCLUSIONS

Modern interactions between nature and society and the current types of environmental problems offer new possibilities and incentives for the linkage of natural and social sciences. Ecology as a natural science is less and less able to observe processes in the biosphere independently of human activities. Environmental problems not only present an opportunity for expanding the sphere of sociological and social research in general, but also represent a theoretical/paradigmatic challenge for the social sciences (economics, sociology, ethics). There are considerable theoretical/methodological difficulties in linking the physical and social descriptions of environmental problems. A more demanding level of unity of natural and social sciences, which is not merely linkage and coordination, establishes itself through such notions as system, evolution, entropy, chaos, catastrophe, etc. Numerous interesting, social/ecological questions cannot be answered by means of individual empirical studies, but only through theoretical reflection and reconstruction of different empirical sources and data. This type of research will in particular demand a new ecological paradigm. Neither the biological/naturalistic nor the strictly endogenous social position contains the epistemological potential for a new ecological paradigm. In this respect Marx's social theory has certain advantages over Spencer's and Durkheim's tradition, yet in some aspects it is ambivalent and even anti-ecological. There exists not only an intellectual and epistemological need, but also a praxeological need for certain levels of unity and holism in science. Fragmentary and disciplinary science also implies fragmentary and disciplinary practice. Ecology has explicitly shown the holistic character of the biosphere and ecosystems, which requires a holistic and interdisciplinary way of thinking. The theoretical formulation and practical realization of the concept of an ecologically sustainable society and ecologically sustainable development will demand and encourage not only multidisciplinarity, but also interdisciplinarity in science, especially as regards the social, natural and technical sciences.

REFERENCES

Barnett, S.A. (1990). 'The Reductionist Imperative and the Nature of Humanity'. *Interdisciplinary Science Reviews* 15:119-132.

Briskman, L. (1987). 'Three Views Concerning the Unity of Science'. In: G. Radnitzky (ed.), *Centripetal Forces in the Science*. Vol. 1. New York: Paragon House Publishers, p. 105-129.

Catton, W.R. Jr. and R.E. Dunlap. (1980). 'A New Ecological Paradigm for Post-Exuberant Sociology'. *American Behavioral Scientist* 24:15-47.

Chen, R.S., E. Boulding and S.H. Schneider. (eds.). (1983). *Social Science Research and Climatic Change*. Dordrecht, Boston, Lancaster: D. Reidel Publishing Company.

Clark, W.C. and R.A. Munn. (eds.). (1986). *Sustainable Development of the Biosphere*. Cambridge, London, New York, New Rochelle, Melbourne, Sydney: IIASA, Cambridge University Press.

Cohen, B.I. (ed.). (1994). *Centripetal Forces in the Sciences*. Vol.1, 1987, vol.2, 1988. New York: Paragon House Publishers.

Commoner, B. (1990). *Making Peace With the Planet*. New York: Pantheon Books.

Daly, H.E. and J.B. Cobb Jr. (1989). *For the Common Good*. Boston: Beacon Press

Dickens, P. (1992). *Society and Nature: Towards a Green Social Theory*. Wheatsheaf: Harvester.

Giddens, A. (1976). *New Rules of Sociological Method: A Positive Critique of Interpretative Sociologies*. New York: Basic Books Inc.

Giddens, A. (1987). *Social Theory and Modern Sociology*. Cambridge: Polity Press.

'Global Environmental Change' (1991). *International Social Science Journal* 43:599-669, 707-718.

Hampicke, U. (1992). *Ökologische Ökonomie*. Opladen: Westdeutscher Verlag.

Kahn, A.E. (1966). 'The Tyranny of Small Decisions: Market Failures, Imperfections and the Limits of Economics'. *Kyklos* 29:23-47.

Kocka, J. (Hrsg.). (1987). *Interdisziplinarität*. Frankfurt am Main: Suhrkamp.

Kuik, O. and H. Verbrunggen. (eds.). (1991). *In Search of Indicators of Sustainable Development*. Dordrecht, Boston, London: Kluwer Academic Publishers.

Locke, J. (1964). *Two Treatises of Government*. Cambridge: Cambridge University Press.

Limoges, C. (1994). 'Milne-Edward, Darwin, Durkheim and the Division of Labour: A Case Study in Reciprocal, Conceptual Exchanges between the Social and the Natural Sciences'. In: B.I. Cohen (ed.), *The Natural Sciences and the Social Sciences*. Dordrecht, Boston, London: Kluwer Academic Publishers, p. 317-345.

Machlup, F. (1969). 'On the Alleged Inferiority of the Social Sciences'. In: L.I. Krimerman (ed.), *The Nature and Scope of Social Science*. New York: Appleton-Century-Crofts, Educational Division, Meredith Corporation.

Mirowski, P. (1989). *More Heat than Light. Economics as Social Physics: Physics as Nature's Economics*. Cambridge, New York, Port Chester, Melbourne, Sydney: Cambridge University Press.

Nagel, E. (1961). *The Structure of Science*. New York Burlingame: Harcourt, Brace World Inc.

Odum, W. E. (1982). 'Environmental Degradation and the Tyranny of Small Decisions'. *BioScience* 32:728-729.

'Reconciling the Sociosphere and the Biosphere' (1989). *International Social Science Journal* 41:297-449.

Robinson, J.B. (1991). 'Modelling the Interactions between Human and Natural Systems'. *International Social Science Journal* 130:629-649.

Roegen, G.N. (1971). *The Entropy Law and the Economic Process*. Cambridge, Massachusetts, London: Harvard University Press.

Roth, G. (1992). 'Autopoesie und Kognition. Die Theorie H.R. Maturana und die Notwendigkeit ihrer Weiterentwicklung'. In: S.J. Schmidt (ed.), *Der Diskurs des radikalen Konstruktivismus*. 5. Auflage. Frankfurt am Main: Suhrkamp.

Saggoff, M. (1980). 'On Integrating the Environmental Science'. In: *Sustainable Development, Science and Policy*. Oslo: The Norwegian Research Council for Science and the Humanities, p. 377-397.

Schlegel, R. (1977). 'Why Can Science Lead to a Malevolent Technology?'. *The Centennial Review* 21:18.

Sorokin, P. A. (1957). *Social and Cultural Dynamics*. Vol.1. Boston: Porter Sargent.

Turner, J.H. (1991). *The Structure of Sociological Evolution*. Fifth Edition. Belmont: Wadsworth Publishing Company.

Varela, J.F. and J.P. Dupuy (ed.). (1992). *Understanding Origins. Contemporary Views on the Origin of Life, Mind and Society*. Dordrecht, Boston, London: Kluwer Academic Publishers.

NOTES

1. Environmental problems are a specific field and an opportunity for linking natural and social sciences. Odum gave his work "Ecology" (1975) the subtitle "Link between the Natural and the Social Sciences". In a once purely biological ecology, the influence of human activity on the bisphere is gaining a central position. The era in which ecology observed the processes and changes in the biosphere completely independently of human activity is on decline. Ecology as a natural science is changing into human ecology in this sense, too. Francois Ramade gave his work "Elements d'ecologie applique" (1978) the subtitle "Action de l'homme sur la biosphere". Numerous ecologists have made contact with economic, ethical, political and sociological problems and, vice versa, numerous sociologists have begun to discover the social causes and social dimensions of environmental problems. They saw the implications for social theory and practice in ecological findings and facts. In these processes one might even recognize a concretization of Marx's vision of science, that the science of human beings will become the science of nature and vice-versa.

2. This is well documented in I. Bernard Cohen's (ed.) "The Natural Sciences and the Social Sciences", 1994.

3. Daly and Cobb (1989: 33) give an example of the split between economic and demographic growth, which was once a common theme within economics itself. Later on this relation was neglected by economics, as the population became the subject of an independent discipline. For economics, the relation between the population and the economy became something external, which is also what happened to demography. The once inherent connection of two themes within a certain science became something external and marginal for both disciplines. Linkage is now possible only in an interdisciplinary context. Although the transformation of an intradisciplinary relation of certain aspects of reality into an external interdisciplinary relation is typical for a process of specialization, all epistemological consequences of this process are not enough investigated.

4. Gerhard Roth (1992) stressed that the theory of social behaviour, the social system, must consider the difference between autopoiesis and cognition because only cognitive systems are operationally closed, and because social systems are not autopoietical in the same sense as biological systems are. Such autopoiesis is, for Roth, reserved only for biological systems. Cognitive and social systems are self-referential, not autopoietical. From the ecological aspect it is important for biological and social systems that they are open as regards material and energy flows. For this reason there are no fully autonomous systems, but only relatively autonomous ones. Furthermore, social systems which include technology are also specific entropic systems. Roth called attention to an important feature of cognitive systems, namely that they were able to develop because they had managed to overcome their subordination to the autopoiesis of the system. Thanks to their attained freedom they will be able to serve autopoiesis. In my opinion the production of scientific findings, technological and social innovations, inventions of new needs and ways of fulfilling them are the most evident manifestation of such service. But isn't the freedom of

cognition against autopoiesis the very source of environmental problems? Within the frame of this freedom, cognition develops to the point where degradative environmental processes begin to endanger the human organism.

5. 'The fish is not only, existentially, a fish, but also an element of this network, which defines its functions, indeed, in the evolutionary sense, a good part of the network - the microorganisms and plants, for example - preceded the fish, which could establish itself only because it fitted properly into the preexisting system.
In the technosphere, the component parts - the thousands of different man-made objects - have a very different relation to their surroundings. A car, for example, imposes itself on the neighbourhood rather than being defined by it.' (Commoner, 1990: 8).

4. The Establishment of an International Regime: the Case of the Stratospheric Ozone Depletion

August Gijswijt

Abstract[1]

The ozone layer protects life on earth against the ultraviolet radiation of the sun. Since attention was first drawn to the problem in 1974, this question has gone through a complete policy cycle at the global level. In 1985 an outline treaty was concluded in Vienna, which was elaborated in 1987 in the Montreal Protocol and further fleshed out in 1990 in London and 1992 in Copenhagen. The treaty represents a remarkable example of international coordination of environmental policy. With the establishment of the "ozone regime" a large number of countries accepted in the framework of the UN a drastic restriction on the production of CFCs and halons, chemical substances with a number of lucrative industrial uses. Does the success of the ozone regime herald more forceful international environmental policy? Or has the international community defined it as a unique problem, for which a unique solution had to be found?

1. INTRODUCTION

In January 1972 the radiochemist Sherry Rowland of the University of California attended a meeting of the Atomic Energy Commission in Fort Lauderdale. He learned that the British chemist James Lovelock had traced light concentrations of chlorofluorocarbons (CFCs) in the atmosphere which he wanted to use for the observation of air currents. Rowland was struck by the fact that these concentrations roughly corresponded to the total production of CFCs up to that time. Normally speaking certain chemical reactions, rain and gravity remove chemical substances from the atmosphere. CFCs, however, are extremely stable and moreover relatively insoluble in water. Evidently the atmosphere functions as a reservoir for CFCs. Because CFCs are so stable, Rowland saw no immediate problems and let the matter rest for over a year. He did then put a young, recently recruited researcher, Mario Molina, to work on the scientifically interesting phenomenon of the accumulation of CFCs in the atmosphere. Rowland and Molina figured out that the CFCs ultimately reached the stratosphere in and above the ozone layer and were broken down there by ultraviolet radiation, whereby atomic chlorine was released. Molina then made a spine-chilling discovery. He found that chlorine as a catalytic agent broke ozone down into ordinary oxygen. One chlorine atom was capable of destroying 100,000 molecules of ozone. This data, plus the statistics on the world production of CFCs, led to the conclusion that within the foreseeable future the ozone layer would be depleted. Rowland and Molina informed colleagues and submitted an article to *Nature*. Five months later, on 28 June 1974, it appeared in print. Thus the case of the stratospheric ozone depletion arose almost by chance. After attention had been drawn to the ozone problem scientists, environmental movements and other population groups in a number of countries managed to get governments under the umbrella of the United Nations to take globally applicable measures to end the stratospheric ozone depletion[2]. In 1985 an international outline treaty was concluded in Vienna, which was elaborated in 1987 in the influential Protocol of Montreal and further fleshed out in 1990 in London and in 1992 in Copenhagen. Agreement has been reached between all the producer and most of the consumer countries on discontinuation or drastic reduction of the production and consumption of ozone-depleting substances. The treaty contains rules of procedure for incorporating new information into policy. The treaty is now regarded as a classic example of successful diplomacy, which has led to a global regime (Gupta, 1993) for dealing with an important environmental problem (Benedick, 1991:1-8 and 210-211).

In my analysis of the discussions which have taken place, I devote attention to the formation of opinion on the ozone question in some of the countries concerned - in particular the USA, as well as England and The Netherlands. I have chosen scientific circles, the industry concerned and the environmentalists as opposing pressure groups, amongst the public and within the political and administrative gremia. The central themes are: on the basis of what knowledge did people arrive at definitions of the problem, what were the "production conditions" for the acquisition of that knowledge and what were the consequences in terms of accepting responsibility and setting policy in motion which were attached to those actions. How was a new definition of the problem propagated, which provided opportunities for international consultation and led to the formation of a global regime? What conflicts were there in this context - who opted for the new definition of the problem, who were against, and what arguments and weapons were available to the various parties?

2. RAPIDLY INCREASING PROBLEM AWARENESS; EXPANSION AND DIFFERENTIATION OF KNOWLEDGE (1974-1980)

In the first instance the industry waved the article of Rowland and Molina aside. There was scepticism about the accuracy of the theory, but there was an immediate attempt to disqualify the bearer of the news and to undermine his credibility. In 1974 there was a congress of the American Chemical Society at which Rowland was to present a paper on the question. Someone from the chemical concern DuPont tried in vain to prevent the presentation[3]. During his presentation Rowland warned of the danger of an increase in skin cancer as a result of the depletion of the ozone layer. Although many colleagues were still sceptical, the news made the headlines of the world press. Scientific teams from Michigan and Harvard supported Rowland and this led the American administration to collect more information via a committee of the prestigious National Academy of Sciences (NAS). Rowland was appointed a member of the committee. Consumer organizations and local authorities in the towns where the scientists were living immediately initiated actions. The movement spread over America and the sales of aerosols began to decline. It was evidently a question of a substantial shock effect on American public opinion. As far as the industry was concerned, it was no small matter. In 1973 more than two billion air sprays with CFCs were sold in the USA. The industries which were directly involved in CFC production had a global turnover of eight billion dollars in 1974 and provided work for 200,000 people. On 11 December 1974 the first of the innumerable hearings began which the American parliament has devoted to the ozone question. The industry put forward the argument that: 'We may be about to extrapolate an unproven speculation, one open to serious question, into conclusions and laws that could disrupt our economy and indeed our way of life' (Roan, 1989:38) A representative of DuPont did, however, promise that the production of CFCs would be stopped if the theory proved correct (Benedick, 1991:12). The environmental movement formed the National Resources Defence Council as its lobby in Washington. The industry established the Alliance for Responsible CFC Policy with the same objective.

The ozone question came into the limelight worldwide. In The Netherlands on 4 October 1974 the left-of-centre daily *de Volkskrant* reported on the findings of Rowland and Molina. On 3 December, five months after the publication in *Nature*, two MPs of a small left-wing party (PPR) put the first questions about the ozone layer in the Second Chamber of the Dutch parliament (Gijswijt and Van der Vliet, 1993:14-19). The American Administration banned the use of CFCs in aerosol sprays in 1978, thus being far ahead of the greater part of the international community. The basic idea behind this policy was that a reasonable expectation of harm is sufficient cause to intervene. This affected a 3 billion dollar turnover, after, as already mentioned, the consumers themselves had already accounted for a substantial decline in the sale of aerosols. The industry immediately started up research on substitutes. Canada, as a small producer, and Norway and Sweden as exclusively consumer countries, had already followed Amercian policy and introduced a similar ban. The European Community observed this with amazement. 'This approach was (...) destined to perplex EC negotiators throughout the subsequent international debate' (Benedick, 1991:24).

In The Netherlands on 3 February 1978 the PPR again put questions about the ozone question and urged that measures should be taken. Since there was no support from other parties for their standpoint, the government was able to get away with vague promises about an intended 50% reduction of CFCs in aerosol sprays (Gijswijt and Van der Vliet, 1993:22-23). In 1977 The Netherlands had indeed proposed in the EC that labels on aerosol sprays should mention the use of CFCs. In 1979 West Germany proposed a ban on the use of CFCs in aerosol sprays in the EC. Italy, France and England, three countries with a considerable production and consumption of CFCs and lacking an environmental movement alive to the situation, blocked both proposals. The EC standpoint was principally determined by Atochem and ICI, the two big producers of CFCs in respectively France and England (Benedick, 1991:32-34; Haas, 1991:231; Jachtenfuchs, 1990:265, 268 and 275). Up to 1975 the world production of CFCs rose by 16% per annum, followed by a decrease under the influence of the action of the American public and the American administration. In Europe too actions by and amongst consumers led to a drop in the sale of aerosol sprays with CFCs. As a result of the enormous attention to the ozone question in the USA research was greatly stimulated. Scientific knowledge increased considerably, but also became more complicated. Prognoses concerning the depletion of the ozone layer began to vary substantially[4]. This made it more difficult to arrive at unequivocal interpretations of the ozone question. This differentiation of knowledge had a negative influence on the causal perception of the problem amongst the population, the pressure groups concerned and the government.

3. FADING CONCERN; EMPIRICAL RESEARCH AS A SELECTION MECHANISM FOR THE ACCURACY OF RIVAL THEORIES AND HYPOTHESES (1981-1986)

In 1981 the newly elected president Reagan appointed Anne M. Burford as director of the Environmental Protection Agency (EPA). Burford blocked every step towards progress in the international discussion on the ozone question. The CFC industry ceased its efforts to develop substitutes for CFCs. New possibilities of using CFCs emerged in the packing industry and in the production of cars and consumer electronics and world production rose in 1982 to regain the level of 1974. In 1983 and 1984 production increased still further (Haas, 1991:226-227). The industry in America feared that market-regulating measures would only apply to the USA. In so far as the American administration initiated policy to restrict production, the aim was to do so in the international framework, just as there had been an attempt earlier on to harmonize policy within America (Benedick, 1991:31). In 1983 Burford was replaced as director of the EPC by William Ruckelshaus. 1983 also saw the formation of the Toronto group of countries (Canada, Finland, Norway and Sweden), which proposed a worldwide ban on the use of CFCs in aerosol sprays. The EPA now got the USA to follow this course of action.

The malaise in the international consultations on the ozone question lasted until mid-1986. Thanks to the driving force of the United Nations Environmental Programme (UNEP) with its secretary-general Mostafa Tolba[5] an outline treaty was achieved in March 1985 during a conference in Vienna. As far as content was concerned, it did not amount to much, but it did form the basis for the much more far-reaching Montreal Protocol in 1987. In the USA the prejudiced attitude of the Reagan administration formed a real obstacle. 'The President's cabinet seemed to have utmost confidence that science would prevail and save the world from any peril caused by chemicals'. Reports were dismissed with catch phrases such as: people form the 'ultimate resource' and are so resourceful that they will certainly solve the problem in the future (Roan, 1989:115). Interest in tackling the ozone problem revived when measurements of the thickness of the ozone layer in Antarctica became available, from which it was evident that there was unexpectedly severe depletion (Roan, 1989:125-141; Gribbin, 1992:81-84). It is once more striking to what extent chance and the perseverance of an individual were decisive in making fresh knowledge available. As early as 1957 the Briton Joe Farman had had a limited and inexpensive program running to observe the ozone layer in Antarctica, which he maintained with the greatest difficulty. In May 1985 he and his team published their findings in *Nature*. The laboratory theory of Rowland and Molina predicted a slow decrease in the amount of ozone. Farman on the other hand observed a strong decrease in the spring[6]. Much earlier NASA had carried out advanced satellite observations, but appeared not to have 'seen the hole in the ozone layer', because the computer automatically replaced low values in the series of measurements by a standard value of 180 Dobson units[7]. Lower values were considered out of the question. Perhaps the fact that the satellite could not take measurements during the polar night played a role. The original data were still available for further analysis and confirmed Farman's observations. This case is a clear example of the social construction of knowledge, whereby measurement data, which are considered impossible, can plausibly be argued away. The observations of Farman and the NASA caused a considerable stir in scientific circles. A whole series of explanations for 'the hole in the ozone layer' above Antarctica were put forward. The original premise of Rowland and Molina was a chemical theory. A 'dynamic' theory on the replacement of air masses with a high ozone content by those with a low content as the cause of the decrease was the principal rival, together with a theory about the role of nitrogen dioxide under the influence of the increase and decrease in UV radiation in the eleven year solar cycle. Rowland assumed that the surface of particles such as dust (from volcanos) and/or ice-crystals accelerated the reaction between chlorine and ozone. In the autumn of 1985 a more complete theory was presented, in which the existence of ice clouds up to a great height in the Antarctic stratosphere played a crucial part. In August-October 1986 the *First Ozone Expedition* to Antarctica, organized by the National Science Foundation, took place. The nitrogen dioxide theory was negated by the observations. In the November

1987 number of *Science* Molina and colleagues provided definite proof of the ice-crystal theory with laboratory tests. Computer color simulations of changing ozone concentrations above the South Pole were frequently shown on TV and reproduced in periodicals and daily papers. Public interest in the question was again greatly on the increase. NASA and the NSF quickly found a budget for a *Second Ozone Expedition* in August 1987. Both expeditions were indeed partly financed by the Chemical Manufacturors Asocation in the USA, who thereby kept a certain distance from the CFC industry. At the same time an Ozone Trends Panel was set up within the World Metereological Organization (WMO) at the instigation of the NASA expert Bob Watson. This consisted of some 150 experts from all over the world. This was the proper method for transmitting messages to other governments as objectively as possible and backed by the weight of international scientific consensus. On the grounds of the observations above Antarctica the American organization of CFC producers (Alliance for Responsible CFC Policy) declared on 16 September 1986 that it was desirable to set a limit to the production of CFCs. The new president of the Alliance, Barnet, described as more 'open-minded' than his predecessor, stated that the scientists had not been able to demonstrate any actual risks, but were not in a position to exclude them entirely (Roan, 1989:192). With this pronouncement he reversed the onus of proof; an unusual standpoint to be taken by the industry in the ozone discussions. A few days later DuPont declared that they would support a further worldwide restriction on production, as proposed by the UNEP (United Nations Environmental Program). Research was also restarted on substitute substances for CFCs. DuPont researchers had tested the theory of Rowland and Molina with their own models and now aligned themselves with the Ozone Trends Panel. It was no great sacrifice: some 2% of DuPont's turnover and a good 2% of their profits came from the production of CFCs, and the market for substitute products was wide open. This manoeuvre by DuPont had important consequences for the determination of the standpoint of the European industry (Haas, 1991:231).

4.   THE REALIZATION OF THE OZONE REGIME (1986-1987)

By virtue of the Clean Air Act the USA Minister of State was empowered to carry on negotiations on behalf of the President concerning an international treaty on the ozone question. In the meantime the active coalition of the State Department - Secretary of State Schultz was receptive to the ozone problem - and the Environmental Protection Agency (EPA) formulated a new standpoint for entering into international negotiations. In the short term they wanted to freeze the production level of fully halogenated CFCs, in the long term to stop production altogether and to review policy periodically if new data gave reason to do so. The relatively influential Natural Resources Defence Council (NRDC), formed in 1974 on the initiative of the environmental movement, advocated a 30% decrease in use in 18 months and a complete cessation of production over a period of 10 years. This tightening up was accepted almost in its entirety (95% reduction in 10 years) by the American administration. In December 1986 the Americans put forward this proposal at the UNEP conference in Geneva on the international approach to tackling the ozone question. The EC and Japan were against. France, Italy and England blocked attempts by Denmark, Germany, Belgium and The Netherlands to accept a vigorous policy line for the EC. Doninger of the NRDC had the impression as an observer that Europe and Japan still had no real idea of the seriousness of the situation (Roan, 1989:196). The scientific lobby in Europe did not carry much weight. The British government, for instance, was not prepared to allocate Farman £35,000 to enable him to augment his technical observation equipment. Ultimately the American Chemical Manufacturors Association provided a budget for the British researchers from a research program on CFCs, which amounted to 18.9 million dollars in the period 1972-1985 (Haas, 1991:226). The American senators Baucus and Chafee were disgusted with the line taken by important countries, such as Japan and the Soviet Union, and the EC. They wanted to introduce legislation in the USA which would put a stop to production at a fixed time. The CFC industry was frightened of a solo by the USA and alerted the Reagan administration. Resistance against the policy of the EPA and the State Department was growing in the financial-economic corner of the administration. At a meeting in the White House Interior Secretary Hodel suggested a plan to decrease exposure to UV radiation by 'personal protection and change of life style'. The ideal of people's maximum personal responsibility for what happens to them, in combination with an unresponsive administration, culminated here in a caricature. The jeers in the media and the disapproving reactions from public opinion resulted a few weeks later in the Senate adopting by 80 votes to 2 a resolution, which stated that in the international negotiations the USA should proceed on the assumption of an immediate freeze of production and consumption levels, of a reduction of 50% in the short term and an ultimate ban on the production of CFCs. The White House had to climb down (Benedick, 1991:59-61). However, the American Office of Management and Budget continued to oppose a strong stand in the forthcoming international negotiations in Montreal. This office now took the line of complicating the USA negotiating position by drawing up unrealistically severe conditions for an agreement. At the same time, but contrary to this, the State Department opened a world wide diplomatic offensive in favor of regulation, concentrating in particular on Japan and the Soviet Union, and on Belgium, which was to take on the chairmanship of the EC in the first half of 1987. Running parallel to this the EPA organized a large number of international scientific meetings. In America the EPA launched a campaign directed to the media and the public. This offensive was a success, as was later to become apparent. President Reagan settled the controversy within his administration in favor of the State Department and the EPA, but for tactical reasons kept it very quiet (Benedick, 1991:65-67). A strong policy could in any case be defended if only on economic grounds, by weighing the advantages of continuing production of CFCs against the costs of increasing cases of skin cancer (estimated at 1.4 million in forty years), considerable damage to cotton and grain on account of UV radiation and the more rapid obsolescence of synthetics (Roan, 1989:204).

In April 1987 there was an important meeting of scientific experts as part of the preparations for the big UN conference in Montreal. Their models supported the view that international agreements on firm policy were a vital necessity. They also showed that there were more harmful substances than CFCs alone. This was based on a measurement of a substance's ozone depleting effect: the 'ozone depleting potential' (ODP). At the same time the governments in a growing number of countries showed increasing readiness to try to achieve a worldwide ozone regime. Though the attitude of the EC with regard to regulation remained negative, a number of member states - Belgium, Denmark, The Netherlands and West Germany - now openly distantiated themselves from the EC position. In September 1987 representatives of more than sixty countries, over half of them developing countries, met in Montreal. In contrast to the 1985 conference in Vienna, the environmental movement clearly made its presence felt. The other pressure group, the CFC industry, was as always well represented, as were the media (Benedick, 1991:74). The USA demanded that an agreement should not be valid until the countries jointly responsible for 90% of the consumption of CFCs had appended their signature[8]. The Soviet Union with 10% consumption and Japan with 13% would hereby be given the opportunity to veto. This was a heritage from the internal American controversy and the last, in itself crafty, attempt by the opponents of regulation to make an agreement impossible. Confusion arose and failure threatened. In the compromise achieved the 90% was changed to 67%. A number of important countries, including the principal EC member states (just as the EC itself) and Japan, signed the agreement, in which it was decided to freeze production and consumption in 1990 at the level of 1986. A reduction of 20% was agreed for 1 January 1994 with a further reduction of 30% in 1999. The production of halons would be frozen in the mid-nineties. A fund was established for the introduction of new technology for CFC substitutes in developing countries. An important item was the agreement ultimately to arrive at a production stop. It was also agreed that the UNEP could convene the participating countries on the grounds of new information. For the first time since 1978 concrete progress had been achieved. There were, however, a number of saving clauses. Thus developing countries were allowed to slightly increase their low level of consumption.

To sum up: mid-1986 saw the start of the crucial phase which led, surprisingly in view of the difficult situation at that point in time, to the result of Montreal in September 1987, namely that agreement was achieved. Important factors were:

- the stand taken in the USA itself, which involved a hefty controversy between opponents and supporters of regulation; as a result the international position of the USA was implausible up to 1987; the pro-regulation faction won this conflict and was also very active on the diplomatic front;

- the stand taken within the EC; the European Commission was totally tied to the apron strings of the industry in France and England until a rift occurred within the EC;

- the attitude of the Soviet Union; glasnost was beginning to exert influence on the Kremlin's international actions;

- Japan; after initial opposition the government came round to the American point of view;

- the situation of the third world countries; in view of their low consumption they were (temporarily) given an exceptional position.

Central themes in Montreal; the difference in opinion between the USA and the EC

Benedick (1991:77-97) names seven themes on which the discussions concentrated in Montreal and in the lead up. With each theme the difference between the USA and the EC was manifest. The most important controversy concerned the status of the EC versus that of the member states. With an eye on Maastricht 1992 the EC wanted to be treated as one party, but could not guarantee that all the member states would join in. Greece, Portugal and Ireland were even absent in Montreal. A number of countries, including the USA, feared that heavily reducing countries such as West Germany would free capacity for other countries within the EC. In the last stage of Montreal everything threatened to go amiss on this point. The New Zealand Minister of the Environment suggested the compromise that the EC countries should adhere *jointly* to one ceiling as regards consumption and one ceiling per country for the production. All the EC countries separately plus the EC itself should ratify the treaty. This compromise was accepted. The USA ratified the treaty in March 1988 as one of the first countries. Shortly before the deadline of 1 January 1989 - which, had it been exceeded, would have meant a nine months delay in the protocol becoming operative - the EC ratified. In the formation of the international regime the USA functioned as a political unity with a national public opinion, with a coordinated environmental movement and with a prominent and powerful scientific community. The EC also functioned as a political unity, but without a supranational public opinion embracing all the member states, without media which covered the whole community, without a coordinated environmental movement and without a powerful scientific forum which penetrated to Brussels and Strasbourg. It became evident here just how weak the institutionalization of political democracy was in the European Community. In the ozone question the European Commission, as has been said, functioned up to 1987 predominantly as the advocate of the interests of two major CFC producers: ICI in England and Atochem in France, backed up by Italy and not hindered by the other Southern member states.

5. TIGHTENING UP OF THE OZONE REGIME (1987-1992)

In the meantime the second ozone expedition to Antarctica was in progress. There were now two planes available with special apparatus, one of which carried out observations at fairly great risk at a very high altitude in extremely low temperatures. Incontrovertible data was acquired on the correlation between the increase of chlorine monoxide and the depletion of the ozone layer. Simultaneously Farman demonstrated that the drop in temperature in Antarctica was responsible for changing chlorine from a non-active stage into an active one, and Molina published his previously mentioned spectacular article in *Science*. In the same number of the periodical another researcher published similar results which had been obtained with the aid of a different type of laboratory experiment. On 1 December 1987 the American senator Albert Gore[9], nominated as presidential candidate for the Democratic Party, brought up the ozone question in a prime-time TV debate with other candidates. On 14 March 1988 the Senate ratified the Montreal Treaty by 83 to 0. The same day the Ozone Trends Panel produced a report mentioning a worldwide loss of ozone. 'The panel's conclusions made headlines around the world. Ozone layer depletion was no longer a theory; at last it had been substantiated by hard evidence. And CFCs and halons were now implicated beyond reasonable doubt'. (Benedick, 1991:110). Four days later DuPont decided to stop the productions of CFCs as soon as substitutes were available. In England the same pattern of reaction occurred somewhat later (Maxwell and Weiner, 1993:28-36). In August 1987 the UK Stratospheric Ozone Group published an overview of the scientific situation. Though international views were well reflected in the content of the report, the conclusion, written by the Department of the Environment, emphasized that it was not really possible to say anything about the causes of the hole in the ozone layer. Nevertheless the government, industry and the public in England now really reacted with concern. In the mid-eighties Thatcher had still been extremely negative about the environmental groups and had called them 'the enemy within'. But in September 1988 she held a speech for the Royal Society, in which she emphasized that a healthy economy and a healthy environment are closely connected and devoted specific attention to the ozone question. Prince Charles had already earlier expressed criticism of CFCs in aerosol sprays - the principal application in England - and both Houses of Parliament were, under the influence of the agitation amongst the environmental movement, the media and the population, critical of the bad reputation which the government had earned for the United Kingdom in this matter. ICI lifted its opposition, in contrast to Atochem and the French government (Benedick, 1991:115)[10]. At about the same time - in the winter of 1988-89 - during an ozone expedition to the North Pole by German, English, Norwegian and American researchers, it was observed that the concentrations of chlorine compounds in the atmosphere were fifty to one hundred times greater than predicted (Benedick, 1991:120). In the summer of 1988 West Germany held a round of parliamentary hearings on the ozone question. These led to the conclusion that more far-reaching measures were needed than those agreed in Montreal. In the period 1987-1989 the ecological question and the ozone problem again figured prominently on the political agenda in The Netherlands (Gijswijt and Van der Vliet, 1993:22). In this context it is not unimportant that there was a flourishing economy in that period. With the change of views in the British government the balance between advocates and opponents of regulation within the EC underwent a reversal. On 2 March 1989 the Council of Ministers of the EC decided to go further than the proposal of the Spanish chairman (reduction of production and consumption of 85% at a fixed time) and to arrive at a total cessation. The differences of opinion between the USA and the EC had disappeared (Skjaerseth, 1992:299). In most of the EC member states the industry stopped using CFCs in aerosol sprays before a stop was enforced. After Montreal further agreements on the reduction and cessation of the production of CFCs and related substances were concluded at the conferences in London in 1990 and Copenhagen in 1992. In the five years between Montreal and Copenhagen a worldwide ozone regime was established. The agreements reached in Montreal needed tightening up, since the model calculations of the depletion of the ozone layer were constantly outdated by new observational data, which increased the concern. The regime acquired a comprehensive and complicated character: the number of participating countries rapidly increased, more and more substances were included in the regulation and the treaty partners entered into increasingly strict agreements[11] on the permissable levels of production and consumption. Financial agreements were made with developing countries - in particular China and India - over allocations from the fund for support during the switchover from CFCs to alternatives. Every regulation on the diminution or discontinuation of production or consumption depends for its success on accurate, valid data, which are provided by the countries concerned. This is still a major problem. The calculation of contributions to, and allocations from the joint fund call for the drawing up of precise procedures and the backing of an administrative apparatus.

All these factors have led to the ozone regime acquiring the character of an international bureaucracy, in which only scientific experts and diplomats can find their way. The economic consequences of the regime appear to be limited; predictions of economic disasters have not proved correct. New technology which could replace the CFCs and halons quickly got underway. The major industries gained access to a new, profitable market of substitute substances, which was effectively protected against the offer of the cheap CFCs. There are continuing negotiations on the regulation of other substances which threaten the ozone layer. The same pattern still emerges: countries who want a diminution or discontinuation of production and use versus countries who, for reasons of short term economic interests, do not or only at a later stage. Of the big countries France in particular is always in the group which opposes regulation.

6. THE BACKGROUND TO SOCIAL REACTIONS TO THE OZONE QUESTION

The literature provides the following explanation for the successful establishment of the ozone regime:

a) The world was totally dependent on natural science for having its attention drawn to the ozone problem. The knowledge of the world climatic system has increased enormously and the achievement of a certain scientific consensus has been decisive in defining the ozone question as a political issue. Haas (1991) speaks of the emergence of an ecological epistemological community, Benedick (1991) of a scientific community, which brought divergent national interests into line through its research. Maxwell and Weiner (1993) on the other hand emphasize that in particular the observation of the hole in the ozone layer in 1986 justified the scientists, although they had no conclusive explanation for it. In England public opinion and the government only became impressed when the ozone layer above the northern hemisphere appeared to show signs of depletion.

b) In the period 1974-1980 consumers considerably reduced the market for CFCs by a boycott encouraged by the environmental movement. This was above all the case in the USA, but also in (parts of) Western Europe.

c) CFCs were not indispensable. Technology for substitutes became available fairly rapidly (Haas, 1991).

d) The chemical industry had considerable interests tied up in CFC production, but was not dependent on it. The number of factories amounted to 17 in 16 countries (Skjearseth, 1992:296). It was a surveyable market, comparatively easy to control. The Montreal Protocol provided for that control, whereby the danger of 'unconcerned' enterprises and countries continuing to produce, consume and possibly export, was considerably reduced (Skjearseth, 1992; Maxwell and Weiner, 1993).

e) The USA, with its strong scientific[12] and economic position of power, has exerted great influence on the achievement of international consensus on policy concerning the ozone question.

f) The depletion of the ozone layer in the stratosphere is a problem which can affect all countries.

g) A cost-benefit analysis of the use of CFCs drawn up by the Reagan administration for the USA produced decisive arguments for discontinuation. An additional advantage of the policy pursued was that CFCs (and halons) also contribute to the greenhouse effect (Skjaerseth, 1992).

h) The UN and within it the UNEP have been of great significance for providing organizational support for international negotiations.

i) Many authors mention the important role played by people in forming individual networks within and between institutions. A limited number of advocates of an ozone regime got to know each other in small personal networks, whereby the circles of scientists, governments and public institutions, industry and the environmental movement became linked with one another.

Benedick (1991), as a top diplomat of the American administration personally involved in the negotiation process, has the most positive opinion about what was achieved in Montreal and thereafter. Other reasonably successful international treaties have been concluded, but Montreal stands out (cf. also Caldwell, 1988; Sussking and Ozawa, 1992). He is optimistic about the exemplary effect of the ozone treaty. All the other authors who have written on the subject share this opinion to a greater or lesser extent. They are, however, less optimistic about the formation of a regime for other global environmental questions, in particular climatic change through the emission of greenhouse gases.

With this summary the question of the whys and wherefores behind the formation of the ozone regime, and other comparable international regimes, is far from having been satisfactorily answered. In the following paragraphs I shall make a number of suggestions, which can contribute to a more systematic explanation.

6.1 Relative autonomy and the international organization of science as conditions for signalling problems

After Rowland and Molina had made their discovery in 1974, they undertook all sorts of actions in a certain order, whereby they unintentionally showed how the political agenda comes about. First and foremost they needed the support of authoritative colleagues, who could confirm the importance of their hypothesis. This authority derived from the scientific top gives access to the media and government circles and provides a certain degree of protection against the heavy criticism to be expected from 'opponents'. Reputation and standing form the carrier for the transmission of the scientific message. Rowland was a renowned research worker, working for an excellent institute and had access to the administrative networks of the NSF, the NAS and NASA. But he turned up with a tiresome message. The reputation of CFCs was precisely that of being exceptionally safe on account of their extreme chemical stability. In order to see through the danger of CFCs, a considerable mental jump was called for. On account of the chemical inertia of CFCs Rowland himself did not foresee any risk from their possibly remaining in the stratosphere. He waited more than a year before calling in an assistant for further research into the stratosphere as a reservoir for CFCs. And the decision to investigate what happened to the CFCs was entirely connected with scientific curiosity and ambition and not at all with concern about the harmful effects. In the circle of colleagues Rowland and Molina found response, but there was also a great deal of scepticism. They were chemists and not meteorologists, and in the latter circles the scepticism was great. The British had a great name in climatological research, and a certain degree of jealousy both between specialisms and also between countries has played a role. In the meantime the theory of Rowland and Molina scarcely stimulated research into the ozone layer, since they only assumed a very gradual decrease in its thickness. More than ten years later, in 1985, Joe Farman published his sensational discovery of the hole in the ozone layer above Antarctica. Just how anxious researchers are about their reputation is evident from the two years' delay which Farman had to permit himself before publication. NASA confirmed Farman's observations via a different method of observation, which led to the matter gaining tremendous scientific momentum. The alternative theories tumbled over one another, but that did not last very long. The second expedition to Antarctica brought science more into line. The 'ecological epistomological community' (Haas, 1991:225-226), which already existed, gained great international weight through the formation of the Ozone Trends Panel set up by Watson (NASA). This provided for unified perception of 'certain facts and causes' and infected governments throughout the world with the virus of this uniform perception. It is also notable that attention to the question was stimulated on several occasions by discoveries which were capable of several interpretations from the scientific point of view, but which caught the imagination of the public and politicians, such as the hole in the ozone layer in Antarctica and the apparent depletion of the ozone layer above the densely populated northern hemisphere, which brought the problem closer to home. In parliamentary democracies the scientific world has a fairly high degree of autonomy, although budgets come principally from the state. On the basis of the literature discussed here it can be concluded that the scope for scientists to operate independently was greater in the USA than in England and possibly still is. The NAS and the NSF stimulated scientific discussions, nationally and internationally, and could even supply a budget for the first Antarctica expedition and were free to launch this. English experts formed important discussion partners for the Americans. Experts from other countries had little to say.

6.2 The struggle between the environmentalists and industry for the support of the public and the government

In the USA the first reactions both from the environmentalists and also the CFC manufacturers in the ozone affair were to a great extent determined by ideology. The environmentalists were quickly convinced of the gravity of the matter. The public was alerted via the media and a consumer boycott of aerosol sprays was started. The boycott also had political significance for the American administration, who in 1978 announced the prohibition of the use of CFCs in aerosol sprays. The industry was forced into developing technology for substitute substances, but stopped doing so in about 1980 because public opinion became less concerned. Activity did not recommence until 1986 after the discovery of the hole in the ozone layer had once more boosted the thermometer of public opinion. In Western Europe a slight, brief reduction in sales - also set in motion by a consumer boycott - was followed by production being increased again, after 1981 even very considerably. Global regulation thus became of great importance to the American industry and this is relected in the policy of the American administration in the second half of the eighties. In Europe the ozone question was possibly pushed off the agenda to some extent on account of the burning issues of acid rain and the Chernobyl disaster (Benedick, 1991:28).

All in all the CFC industry in America and Europe has reacted unresponsively and, also from the commercial point of view, conservatively. There was no question that the major manufacturers such as DuPont in the USA, ICI in England and Atochem in France were really threatened economically by a ban on CFCs. Sometimes, however, people played up the rhetoric of the threat to the country's economy and even to the individual's way of life. Industrial circles often have great influence on governments and politicians and can then impose their definition of a situation because they can provide for employment. They have at least the power to form a considerable *nuisance*. The application of obstruction power has a retarding effect on decision-making, at any rate from the point of view of the party wishing for change. The Reagan administration had a strong ideological orientation to non-intervention in the market and lent the CFC industry a ready ear. When the nasty facts about the effect of CFCs on the ozone layer could no longer be ignored - the scientific division of DuPont endorsed this verdict in 1986 - industry in the USA accepted a decrease in, and thereafter a ban on production. A new, profitable world market emerged for more expensive CFC substitutes.

6.3 Problem awareness, causal perception and willingness to act amongst the population

People in many countries exhibit concern about the environmental question. Surveys show that they only get really worked up about it - show emotional concern, are willing to take action themselves, will accept relevant policy decisions - when an environmental question directly affects them (Midden, 1993:12). When is this the case? Uncertainty about the causal structure of a problem and about the chance of specific cause and effect patterns occurring diminishes people's willingness to act. The group character (Midden, 1993:11) or the social structure of an environmental problem also influences this willingness to act. When the effects of people's own behavior are not demonstrable, because they disappear in the mass, there is less readiness to act. The environmental question in general - many phenomena and causes, everybody is involved - provides few if any points of contact for causal perception and readiness to act (Tellegen and Wolsink, 1992:106). From an enquiry held in The Netherlands (Midden, 1993:11) it appeared that 75% of the respondents consider the greenhouse effect dubious as regards its magnitude and the moment at which it will occur. 60% consider the risks to be vague. People see their own contribution to the problem as being very slight, consider others much more responsible and show a low degree of willingness to act. One would expect that the ozone question would evoke similar reactions. After all, the ozone case has been described as a problem which only becomes manifest in the long term, as a question of substances which people cannot see or smell, which destroy something way above their heads, whereby they become vulnerable to radiation which they also cannot perceive. Nevertheless in 1974 the population of the USA reacted immediately and intensely to the discovery of the ozone problem. The discovery of the hole in the ozone layer above Antarctica and the depletion of the ozone layer - at that point in time scientific enigmas - evoked the same reactions. How is this to be explained? The fierceness of that reaction can be explained through people's personal experience. Almost everyone knows that invisible radiation can be dangerous and can in particular cause cancer. The greenhouse effect is much more indirect and more vaguely connected with people's own health and wellbeing. Furthermore in 1974 the specialized knowledge on the causal mechanisms at issue was very meagre and experts could in all conscience give a simple picture of the state of affairs. Longitudinal research into environmental behavior and environmental awareness in The Netherlands shows that two factual questions on the environmental risk of CFCs in refrigerators and the effect of CFCs in refrigerators were answered correctly by about 77% respectively 67% of the respondents. It is true that these questions are of an elementary nature[13]. When people have only restricted knowledge of a problem, and specialists find it difficult to provide unambiguous interpretations, this gives rise to *unstable problem awareness*. Every crumb of new information about a potential threat can substantially influence the problem awareness. Scientific observations which are still surrounded by uncertainty about their definitive empirical confirmation and interpretation can nevertheless strongly attract the attention of the public, the media and politicians. But conversely observations which can give rise to great concern among the experts may leave the lay public cold.

What had been decisive for the success of the buyer's boycott in America and parts of Western Europe in the seventies was that people felt they had a clear picture of what was the matter. The causal connection between the depletion of the ozone layer and the entry of the CFCs into the stratosphere, as well as their own contribution to this by the purchase of aerosol sprays, was not difficult to grasp. The campaign started in the places where the influential scientists concerned were living. The environmental movement acted coherently and efficiently. Later on the McDonalds concern in the USA was compelled through a consumer boycott to replace wrapping material produced with CFCs by other material. In neither case was it a question of very important consumer articles and alternatives were available. Personal disadvantages from the change of behavior were slight. The uses of CFCs other than in aerosol sprays have seldom been the object of a consumer boycott. Quite a large amount of CFCs is used in the production of cars and consumer electronics. Nowhere has there been an attempt to influence the manufacturers of these products through the consumers. From the point of view of the consumer the use of CFC is only one aspect of the decision to buy. Refrigerators were said to be 'indispensable' and no ecologically sound alternatives were available. The situation has recently altered in the sense that the environmental movement in Germany has managed to get the production of refrigerators started up which work on a mixture of butane and propane as the refrigerant. This gives the consumers the chance to choose, as was the case with aerosol sprays earlier on.

Around 1978 a phase began in which the differentiation in the knowledge and interpretations of experts evoked uncertainty amongst the public, the politicians and the authorities. Even the discoverers of the problem estimated at a given moment that the tempo of the depletion of the ozone layer would be much slower than they had originally calculated. The discovery of the hole in the ozone layer in 1985 and in particular the computer-controlled color simulations of the decreasing and increasing thickness of the ozone layer above the South Pole, really caught the imagination of journalists, authorities and politicians, and of the public. The differing reactions per country to the ozone question are presumably connected with historically developed patterns of sensitivity to the ecological question in general. The role of the media can perhaps partially explain these differences. Research in The Netherlands has shown that quantitative differences in newspaper reporting on the environmental question carry over into the opinions of the readers (Guttling and Caljé, 1993:14-9). Media policy in the various countries can also show systematic differences.

6.4 The attitude of governments

Government often means steering a course between opposing interests and working out compromises. The difficulty with the ozone question is that there is not much scope for compromise. Continuing the production and consumption of ozone-depleting trace gases results in damage which is out of all proportion to the financial and practical objections to discontinuing the use of the chemical substances in question. The attitude most often adopted by governments was that the harmful effect of CFCs should be proved before they were willing to take measures. This is a form of risk perception which can be criticized on principle. In the USA there is an influential current of opinion which wants to reverse the burden of proof if future harm can be assumed. In the initial phase of the international consultations on the ozone problem there was a serious difference of opinion on this point between the USA and the EC. The USA put the question on the international political agenda and was one of the very few countries in a position to do. The *initiative potential* of the USA was based on the availability of scientific expertise, on the relative autonomy of the scientific world - there should be scope for independently making fairly substantial investments - and on a fairly high degree of pluriformity in the views on the environmental question within the (federal) administration. Under Reagan there was a hefty conflict between the staff of the White House and the Office of Management and Budget on the one hand and the EPA and the State Department on the other hand. In Congress there is also a good deal of scope for congressmen and senators to adopt an independent standpoint and work it out. The Senate hearings are repeatedly used to acquire and to evaluate information on the ozone question. During the conflict with the staff of the White House and the Office of Management and Budget, the EPA and the State Department had enough manoeuvring space to carry on with their pro-regulation campaign both at home and abroad. In England under Thatcher this would presumably have been totally impossible. Up to and including Montreal the USA was the driving force behind the establishment of the ozone regime. This was also the case with an international regulation for radioactive waste and combatting the polution of the sea (Liberatore, 1993:14). Under Reagan and Bush the pioneering role of the USA was somewhat reduced. The political color of a government certainly has an influence on the environmental policy. It would not therefore be correct to consider the active policy of the USA on the ozone question as representative for its total international environmental policy[14]. In the formation of the ozone regime Canada and some of the northern European countries played a stimulating role alongside the USA, joined in later phases by Germany and The Netherlands. The attitude of France, Italy and England was averse to a strict regime. Other EC countries took a mainly neutral stand. Such government policy was usually not impeded by criticism from the country's own media and the population. As economic and political superpowers Japan and the former Soviet Union were probably the only other countries with the potential to take the initiative for the establishment of an ozone regime. The EC belongs in this group, but has the disadvantage of laborious internal decision-making procedures. The necessity of taking important decisions unanimously or with a qualified majority gives the member states who want to maintain the status quo considerable power to veto or to hinder. But the EC has also built up a tradition of consulting pressure groups. Extensive consultations on policy conerning the ozone regime were held with the industries involved, and the environmental movement evidently did not take much trouble to intervene (Mazey and Richardson, 1993:123). With regard to the European Commission and the Council of Ministers the European Parliament occupies a weak position.

6.5 Risk perception and cognitive style

The difficulty with which Rowland and Molina and with them the problem-signalling scientists were confronted is that in modern industrial societies there is a fundamental overestimation of the extent to which the risks attached to activities can be fully and adequately assessed by means of scientific analysis. There are four separate dimensions to be distinguished in the observation of relations between humans and nature (Wynne, 1992:115-117):

1) *Risks* within a known system can be calculated. It has now been established that the depletion of the ozone layer leads to a greater chance of skin cancer, cataract, diminished functioning of the immune system in humans and (sometimes) in animals, and to the breakdown of chlorophyl in plants with all the consequences for the production of oxygen in the ocean and on land.

2) There are also *uncertainties* about parameters of a vaguely known system. A great deal is still obscure about what will happen to the ozone layer.

3) *Unknown factors* are much more difficult to deal with. Two chemicals developed in this century, namely DDT and CFCs, were initially received with great enthusiasm. Only much later was their production and use reduced with great difficulty on the grounds of the recognition of effects which had been totally unforeseen. *The greater the spread of scientific application, the greater the domain of ignorance becomes.*

4) Conventional risk analysis treats all uncertainties as belonging to a cause and effect system which is in principle recognizable and to which estimation methods can be applied. The social embedment of technology introduces a type of risk which is based on indeterminate factors which

arise from human actions[15]. The assessment of the risks of nuclear energy was based on presuppositions of a high level of maintenance and management of nuclear power stations. As long as the presuppositions tally, the risk assessments have a certain validity. But uncertain factors such as management blunders and malicious manipulation of (components of) nuclear power stations *cannot be included in a risk assessment*.

For a long time the myth of the invulnerability of nature, combined with a vague but firm faith in people's inventivity, dominated the views on the ozone question of the Thatcher government and a part of the Reagan administration. This line of thought naturally finds a breeding ground in the opinions which are alive amongst the population. This myth forms part of a cognitive style (Haskell, 1985), of almost unlimited optimism about the resilience of nature, and about the scientific resourcefulness of future generations to think up solutions to environmental problems. On the basis of a vague trust in science and technology people are not sufficiently aware of the type of risks which Wynne calls uncertain, unknown and indeterminate factors. The optimism of progress (Achterhuis, 1988) uses the rhetoric of innovation and growth as a remedy for social problems. It combines this with an unconcerned approach to global environmental risks, such as were to be seen, for example, in the ozone question. This abstract confidence evidently provides scope for ignoring the concrete warnings of scientists[16]. There are signs of the emergence of a cognitive style, in which a fundamentally different and much more protective attitude to nature is to the fore. People, or their governments, can make up their mind that extensive areas of what is now still relatively unspoilt nature should remain completely untouched. Antarctica is an example of this, where both the extraction of raw materials and also whaling in the surrounding waters have been prohibited. The realms of ignorance and indefiniteness, which can lead to such unpleasant surprises, are then to some extent restricted. On the basis of a cognitive style, in which restraint is the criterion in dealing with nature, the environmental question loses some of the character of a form of setback.

REFERENCES

Achterhuis, H. (1988). *Het Rijk van de Schaarste. Van Thomas Hobbes tot Michel Foucault*. Baarn: Ambo.

Benedick, R. E. (1991). *Ozone Diplomacy. New Directions in Safeguarding the Planet*. Cambridge, Massachusetts and London: Harvard University Press.

Boudon, R. (1984). *La Place du Désordre. Critique des Théories du Changement Social*. Paris: Presses Universitaires de France.

Caldwell, L. K. (1988). 'Beyond Environmental Diplomacy: The Changing Structure of International Cooperation'. In: John E. Carroll (ed.), *International Environmental Diplomacy*. Cambridge: Cambridge University Press, p. 13-28.

Chunbachi, S. (1985). 'A Special Ozone Observation at Syowa Station, Antarctica from February 1982 to January 1983'. In: C.S. Zerefos and A. Ghazi (eds), *Atmospheric Ozone: Proceedings of the Quadrennial Symposium Held at Halkidiki, Greece 1984*. Massachusetts: Hingham, p. 285-289.

Firor, J. (1992). *Veranderingen in de Atmosfeer*. Amsterdam: Bert Bakker. (Dutch Translation from 'The Changing Atmosphere. A Global Challenge', 1990.)

Gijswijt, A.J., and M. van der Vliet. (1993). 'Regering en Parlement over Mondiale Milieuproblemen'. *Milieu* 8:20-27.

Gore, A. (1992). *Earth in the Balance. Ecology and the Human Spirit*. New York: Plume.

Gribbin, J. (1992). *Om het Behoud van de Ozonlaag. De Rol van Wetenschap, Industrie en Politiek*. Wageningen: Pudoc. (Extended Dutch translation from: 'The Hole in the Sky: Man's Threat to the Ozone Layer', 1988, by P.J.H. Builtjes and D.A. Kraijenhoff van de Leur).

Gupta, J., G. Junne and R. van der Wurf. (1993). *Determinants of Regime Formation*. Working Paper I, Dutch National Research Programme 'Global Air Pollution and Climate Change'. Amsterdam: University of Amsterdam and Vrije Universiteit.

Gutteling, J.M. and J.F. Caljé. (1993). 'De Invloed van het Milieu in het Nieuws: Mondiale Risico's en Risico's Dichter bij Huis'. *Milieu* 8:14-19.

Haas, P. M. (1991). 'Policy Responses to Stratospheric Ozon Depletion'. *Global Environmental Change* 1:224-234.

Haskell, T. L. (1985). 'Capitalism and the Origins of the Humanitarian Sensibility'. *American Historical Review*, Part 1, 90:339-361; Part 2, 90:547-566.

Jachtenfuchs, M. (1990). 'The European Community and the Protection of the Ozone Layer'. *Journal of Common Market Studies* 28:261-277.

Jaeger, C. C. (1993). *Theoretical Perspectives on the Consequences of Climate Change*. Paper 88th Annual Meeting of the American Sociological Association, Miami.

Johnston, H. S. (1992). 'Atmospheric Ozone'. *Annual Review of Physical*

Chemistry 43:1-32.
Kelder, H. (1990). 'Ozon en het Ozongat'. In: H. Tennekes and G.P. Können (eds), *Aanhoudend warm. Klimaatvoorspellingen vanuit De Bilt*. Baarn: Thieme, p. 35-46.
Kempton, W. (1991). 'Lay Perspectives on Global Climate Change'. *Global Environmental Change* 1:183-208.
Kruik, M.D. de, F.G.M. Pieters, W.F. van Raaij and H.W. Mentink. (1993). *Milieugedragsmonitor. Secundaire Analyse van de Derde Meting*. Rotterdam: Erasmus University.
Liberatore, A. (1993). *Beyond the Earth Summit: The European Community Towards Sustainability*. EUI Working Paper No. 93/5, European University Institute, Florence.
Liefferink, J.D., P.D. Lowe and A.P.J. Mol (eds). *European Integration and Environmental Policy*. London/New York, p. 114-25.
Maxwell, J. H. and S. L. Weiner. (1993). 'Green Consciousness or Dollar Diplomacy? The British Response to the Threat of Ozone Depletion'. *International Environmental Affairs* 5:19-41.
Mazey, S. and J. Richardson. (1993). 'EC Policy Making: an Emerging European Policy Style?' In: J.D. Liefferink, P.D. Lowe and A.P.J. Mol (eds), *European Integration and Environmental Policy*. London/New York: Belhaven Press, p. 114-25.
Midden, C.J.H. (1993). 'Milieubedreiging als Gedragsmotief: Cognitieve en Sociale Oorzaken van Onderwaardering'. *Milieu* 8:8-13.
Roan, S. (1989). *Ozone Crisis. The 15-Year Evolution of a Sudden Global Emergency*. New York: John Wiley & Sons, Inc.
Rayner, S. (1993). 'Prospects for CO2 Emissions Reduction Policy in the USA'. *Global Environmental Change*, 3 1:12-31.
Scientific Assessment of Ozone Depletion: 1994. Executive Summary. United Nations Environment Programme, World Meterological Organization, National Aeronautics and Space Administration, National Oceanic and Atmospheric Administration, 19 August 1994.
Skjaerseth, J. B. (1992). 'The "Succesful" Ozone-Layer Negotiations. Are there any lessons to be learned?' *Global Environmental Change* 2:292-300.
Spaargaren, G. and A.P.J. Mol. (1992). 'Sociology, Environment and Modernity. Ecological Modernization as a Theory of Social Change'. *Society and Natural Resources* 5:323-344.
Susskind, L. and C. Ozawa. (1992). 'Negotiating More Effective International Environmental Agreements'. In: Hurrell, Andrew, and Benedict Kingsbury (ed.), *The International Politics of the Environment*. Oxford: Clarendon Press, p. 142-165.
Tellegen, E. and M. Wolsink. (1992). *Milieu en Samenleving. Een Sociologische Inleiding*. Leiden/Antwerpen: Stenfert Kroese.
Wheale, A. (1993). 'Ecological Modernization and the Integration of European Environmental Policy'. In: J.D. Liefferink, P.D. Lowe and A.P.J. Mol (eds), *European Integration and Environmental Policy*. Lon-

don/New York: Belhaven Press, p. 196-216.
Wynne, B. (1992). 'Uncertainty and Environmental Learning. Reconceiving Science and Policy in the Preventative Paradigm'. *Global Environmental Change* 2:111-

NOTES

1. Extensive comments on an earlier version have been made by M. Gijswijt, G.H. Können, T. Korver and N. Wilterdink. I have also made grateful use of comments from B. van Heerikhuizen, H. Kelder, M. Menzel, and J. Sterk.

2. The Protocol of Montreal of September 1987 regulated 8 substances divided into two groups, with a regulation for the addition of other substances to the list, if necessary. The second list, drawn up in London in 1992, already contained 29 substances divided into three groups (Benedick, 1991: 241 and 256-7).

3. Johnston (1992: 29), one of the first top experts who were consulted by Rowland and Molina: 'For the next few years, the industry took the initiative in polarizing the situation and, typically, misrepresented the contents of a scientific article and ridiculed the misrepresented portion'. In a newsletter to the members dated 1 November 1976 the National Academy of Sciences dubbed DuPont's behavior as unworthy of a great institution which had made such an important contribution to chemical research and technological progress.

4. There were prognoses of a decrease in the thickness of the ozone layer from 3 to 19% in fifty to a hundred years (Benedick, 1991: 13).

5. As early as 1975 the UNEP had organized a scientific congress on the American findings concerning the ozone layer. After 1975 the UNEP continued to be very active (Benedick, 1991: 40-50). For the text of the outline treaty of Vienna see Benedick (1991: 218-229).

6. Japanese researchers had already earlier on pointed to low values in the ozone layer above Antarctica (Chubachi, 1985), but Farman woke up the scientific world and alerted public opinion.

7. See Gribbin (1992: 83). The ground measurements are made with a spectograph, which forces out a column of air with ozone into all molecules ozone assembled under a pressure at sea level and at 0 degrees C. One Dobson-unit is then equivalent to a thousandth centimeter ozone. Between 1957 and 1970 Farman and his team measured some 300 Dobson-units in October above Haley Bay in Antarctica. In 1982 the measurements were low, but at that point account was still taken of a possible fault in the measurements. In 1983 and 1984 they came to the conclusion that something was wrong. In October 1987 125 Dobson was measured.

8. This demand was prompted by the desire to lessen the loss of face for the hardliners in the administration, who had just previously had to abandon their resistance to an international treaty (Benedick, 1991: 89).

9. Gore (1992: 3) mentions how his mother told him as a boy in about 1963 about Rachel Carson's warnings.

10. Jachtenfuchs (1990, 275) reports that ICI had in the meantime built up production capacity for CFC substitutes.

11. For the text of the Protocol of Montreal and the London Revisions see Benedick (1991: 230-264).

12. In 1994 the international ozone panel gathered together 293 experts from 36 countries, of which as many as 141 from the USA, 28 from Germany, 26 from England, 14 from France, 7 from Norway, 6 from Japan, 6 from The Netherlands and 5 from Russia (Scientific Assessment, 1994: ii and 13-16).

13. Statement by M.D. de Kruik on the scores for two factual questions. The questions were (Kruik a.o., 1993: XXIX): 1) the cooling systems of refrigerators contain gases which are harmful to the ozone layer; 2) the defrosting of a refrigerator releases CFCs to the air.

14. In a study of emissions of CO2 Rayner (1993: 12) speaks of the 'overwhelming inertia of the US political system'.

15. See Jaeger (1993: 12) and the chapter 'Le determinisme bien tempere´' in Boudon (1984: 165-190).

16. Firor (1990: 106) is head of the research department of the National Center for Atmospheric Research in Boulder, Colorado. His principal argument is that mankind is at the moment carrying out manipulations with acid rain, the depletion of the ozone layer and climatic change which have a greater effect than the large-scale forces of nature. The fact that the consequences of human actions sometimes appear to cancel each other out is not a reassuring thought for the experts. This leads to increased pressure on the system, just as when a patient is simultaneously given medication to check and to stimulate fever.

5. Globalization, Environmental Awareness, and Ecological Behavior Shown at the Example of the Federal Republic of Germany

Wolfgang Schluchter
Andreas Metzner

1. INTRODUCTION

Concerning the environment in Germany, the classical nature conservation idea was prevalent during the turn of the century until the Weimar Republic. Its public articulation of 'life'-movements was permeated by anti industrial romanticism. Its topics were the preservation of landscape, the creation of 'nature monuments', the protection of animals, nudism ('Freikörperkultur') etc. Another topic is the demand of the worker's movement to create housing conditions in line with those of the middle classes, creating the concept of the garden town ('Gartenstadt'). After the devastations brought about by National Socialism and World War II, the Germans were preoccupied with creating their economical miracle ('Wirtschaftswunder'), thus returning to the standards of a western industrial society. Under the influence of a collective economical growth euphoria and an unbroken progression myth, to which big parts of the population were devoted to, the environmental consciousness was fairly non-existent, either limited to mere nature preservation, or motivated by the rejection of dangerous technologies for military projects. But besides the criticism of atomic power of the 'Easter Marches' ('Ostermarsch-Bewegung'), which was more oriented towards peace and opposition against re-arming, there was also a locally defined opposition against further construction of civil atomic plants. The rapid development and spread of environmental consciousness began in the Federal Republic of Germany at the beginning of the seventies. It had two main sources. The first being the social movements, encouraging transformation of growing local and sectorial resistance against the build up of the atomic industry into a general criticism of the prevailing technical-economical oriented narrow growth- and progression-perspectives.
Already the first electoral victories of the Green party made the subject 'environment' a political one and led the established parties to programmatic reactions.
Secondly, the popularization of elaborate global resource-ecological models by the media, was taken up by the administration as a mandate to establish an environmental administration. This would bring the tasks together, that had until now been spread over different departments, to be worked on autonomously.

The report of the 'Club of Rome' in 1972 made widely known the problem of excessive exploitation and destruction of the existential basis of man. It also generated an important stimulus for a change of view of the relationship between man and nature. The idea of an ecological reorientation of economic practice was born, although it met with powerful resistance by the vested interests, even by the unions, who were concerned about the security of jobs and wages. The start and further development of the environmental movement took place in correlation with the so called 'New Social Movements'. Most important was the conflict with technological progress in the form of big atomic generation plants. It was based on various underlying motives. Initially the farmer's and vine-grower's own interests stirred their resistance, against the choice of location for atomic plants. Their concern was obviously about possible harvest and quality losses, e.g. through the emissions of the cooling towers. Yet background motive was also the rural population's fear of cultural foreignization ('Überfremdung') caused by the advance of big industry into the very context of their lives. The next motive evolved, when health hazards were pin-pointed as the central theme by large sections of society. For example radioactive low-radiation, emitted even during normal operation of atomic power plants and the aftereffects of possible accidents, elaborated by critical members of the scientific system and communicated to the citizen's initiatives and action groups. The third motive was fear, that the future would be more and more in the hands of the monopolies. It activated those anti-capitalistic and antiauthoritarian groups, that had formed their attitude during the West-German student's movement. The fourth motive was lack of trust in the scientific experts and representatives of the administration. It was caused by the technicized attitude and belief in progress, prevalent in science and politics, together with its alienation from ordinary understanding and every day conditions. In West-Germany the effects of higher educational standards, generated by the Social Democratic educational reform in 1968, ran along side a decreasing authoritarianism of the better educated social strata. Processes of detachment from handed down dependencies by the then young people of postwar society played another important part, as they induced a search for new identities and social behavior. These processes however, instead of being understood as modernization tendencies and as such integrated into the political arena, they were rejected and vehemently fought against by the established political system. This created a potential of vociferous social criticism, that can be understood as a sounding-board of things obviously gone wrong. The political-administrative system responded by barring a considerable part of the young intelligentsia from careers in civil service professions (by threatening or actually practicing exclusion). It was exactly these personnel, that were involved in the organization of extra-parliamentary, and at the time nonconformist amalgamations of young citizens. These groups spread into the bourgeois strata of society. The connection of the young protesting personnel with established but critical forces was characteristic of the early citizen's initiatives. One could state, that this released considerable social energy and ex-

posed the limitations of the citizen's influence upon the parliamentary-representative system.
As the political system dealt with these movements not in a political-argumentative, but rather in a power-political way, these movements were compelled to analyze the system's underlying interests critically. This not only lead to their realization of its objective contradictions, but also induced a new subjective identity, a modern, self-determined and autonomy striving individual.

2. THE EVOLUTION OF ENVIRONMENTAL AWARENESS AS A LEARNING PROCESS

The awareness of the conjunction and crossing over of local, regional and global ecological problems developed exemplarily in much discussed and widely noticed issues. First it was the correlation of air pollution and the dying of the forests ('Waldsterben'), that received special attention. Later it was the interconnection of air pollution, the hole in the ozone layer, and the greenhouse effect. The disastrous combination of a whole spectrum of causes became apparent - be it the burning of fossil energy through industry and big plants or the emissions of domestic heating systems, be it the multitude of motor vehicles or the effects of large-scale livestock or of individual use of body-care sprays. The Federal Republic of Germany's special location in direct neighborhood to the RGW-States, GDR and Czechoslovakia, where brown coal was burnt on a vast scale with enormous emissions, facilitated at an early stage the realization, that air pollution is a phenomenon across national borders. In the case of water pollution an equal awareness has not yet been developed by the general public. It became apparent however, when due to the poisoning of the Rhine in the aftermath of the Sandoz accident (in 1986), the users of the river water (e.g. water plants and fishermen) had to accept considerable losses and therefore scandalized the incident via their lobbies. The immense response in Germany for the planned sinking of the oil platform 'Brent Spar' by the Shell company shows however, that the pollution of the oceans, especially of the nearby North Sea, is considered generally more and more as problematic.
The increasing globalization of the described phenomena is of great importance for the evolvement of environmental awareness and helps to establish connections between different occurrences. Accidents with devastating aftereffects, like the accident of the Exxon-Valdez, and news about dramatic climate changes, like the desertification of the Zahel-Zone, as well as the consequences of the greenhouse effect, are being more and more connected with each other, so that a pattern of observed ecological, social, and economical interdependencies and conflicts emerges. Within this pattern a split between continued, aggravated erroneous trends and desirable prospects for the future becomes more and more obvious. What indeed is missing, are personalities, who can function as crystallization cores in these processes of development of

new models. Personalities, who are capable of accepting the challenges and can represent them exemplarily, and who can convincingly integrate practical and conceptual knowledge with future oriented outlines. The world-wide expanding importance of organizations like Greenpeace can be explained by the fact, that on one hand they set signals by their actions, and on the other hand they act as emissaries for the silent protest of many individual citizens. The symbolical actions serve as crystallization cores for the ecological-political conflicts. Besides of strategies of problem solving, they appear as limited symbolical politics.

If one looks at the socialization process of individual initiatives and behavior - particularly the principle of fencing off towards the outside while simultaneous forming a group-identity inside - one has to assume, that in view of the global dimensions of anthropogenic produced technical and environmental hazards, the effects of this principle (which constitute the exploitation of the resources and the present way of life) are neutralized. For in the face of dangers of such magnitude, the differentiations within society (e.g. capital and labor) as well as international imbalances develop the tendency to cancel each other out for the simple reason, that all individuals and societies are affected. Yet, some are more affected than others, some are more involved in causing the problem than others, and some are making more profit out of it than others; hence this structural principle does not simply dissolve but gains new dynamics and produces new, critical interpretation models. It makes sense, that the globalization of environmental awareness cannot be a simple matter and cannot be deduced solely from the globalization of ecological problems. It is rather a much more general and complex phenomenon with dynamics, that are propelled also by irregular economical developments. On the one hand international economical connections are expanding and constitute the supposedly necessary compulsion for identical production processes and the conformity of economies. On the other hand this creates a new complexity, that evokes insecurity and fear in the individual. Out of this springs the individual quest for a way into the future and it is motivated differently than the same quest on a social level. It explains the desire of numerous individuals for a possibility to solve problems individually, while at the same time questioning the social mechanisms and competence for problem-solutions. Here the virtually constituted unity of global events, their simultaneity and independency of place, conveyed by the electronic media, plays an important role.

Another aspect of this development is the detectable internationalization of the consumers' consciousness, at least in some highly industrialized countries. In Germany it becomes evident for example in ecologically motivated consumer decisions, like giving up tropical wood, or in politically motivated boycott measures, for example against goods from the former South-Africa. Also the assortment-restrictions of some large trade-chains move into the same direction, when they declare their rejection of Norwegian goods as protest against whaling. Another example is the boycott of the Shell gas-stations in Germany

on the occasion of Shell's planned submersion, which, after loosing 30 % of their sales, must have influenced its decision, to call it off. It can be anticipated, that concerning France's planned atom-test on the Mururoa Atoll, similar campaigns will be launched, which will occur most probably on a much more international scale. Possibly they will show, that towards the end of this century the radius of action of national governments will start shrinking similarly as it did in the case of the multi-national operating Shell-concern. It is inferred, that national actions, carried out without consideration for their social and ecological consequences, will provoke international reactions, which will on the one hand stabilize national organized protest movements inside and on the other hand will globalize them outside.
But these developments can also elicit perverted effects, insofar as the boycott-measures can be passed on as penalty by those, against whom they were directed, towards others, who were not even involved. The intention is to shape a picture of the enemy and shift the guilt towards others. The trend towards internationalism of the 'modernizers' - apparent in boycott measures against countries, whose governments take unwanted, ecological harmful and globally effective steps - can also turn into a formation of 'restorers', as France showed us in connection with the atom-tests in the Pacific.

The fact, that the political Right is gaining strength in all of Europe has to be taken as a warning. The German National Socialism had a slogan: 'The German being shall heal the world'. Also critical ecologically inspired forces in Germany should ergo refrain from telling other nations what to do. What is needed instead, is a criticism, that is accompanied by positive recommendations. It would be good, to propagate the internationalism of the ecological protests and efforts to resolve the problems, according to the internationalism of the early workers' movement.

Closely connected with the growing sensitization for ecological issues in Germany, a democratic change has taken place. During the national phase of restoration in Germany the idea of a representative democracy was an important one. Clearly legitimated and regulated authorities were supposed to function as a guarantor of order and social stability. After the unrest at the end of the sixties, created by the commencement of the students' movement, a decisive change of atmosphere occurred, favorable for a reorientation of foreign policies and educational projects. It was mainly the heavy conflicts between advocates and adversaries of nuclear energy, that broke up the authority of traditional power structures. While the advocates of nuclear energy insisted on the jurisdiction of democratic elected institutions, its adversaries called for more democracy. Concerned citizens demanded competent information and the right to a say in crucial matters.
The establishment argued the old European ideas, originated by Hobbes. They insisted on the principle of 'hypothetical consent': After a government has been elected democratically by the citizens, its decisions are per definitionem

democratically legitimate and the citizens' consent taken for granted. The alternative forces' demand, was the idea of a participation-democracy, where an explicitly given consent to political decisions, granted by informed citizens, is needed. The participation of those involved in public decisions is being demanded. Anti-nuclear action groups of the environmental movement developed together with the New Social Movements various modes of civic influence on governmental decisions, which today - in the middle of the nineties - are an established element of political culture. Nowadays by far more people are involved in citizens' initiatives (about 5 million) than in political parties (about 2 million).

The decision- and development-processes of the present are more than ever characterized by conflict- and consent-oriented public debates. They manifest in collections of signatures, demonstrations, boycott and sit-ins, as well as in hearings or round table discussions.

3. THE RELATIONSHIP BETWEEN ENVIRONMENTAL AWARENESS AND BEHAVIOR

Since the eighties brought environmental surveys and analyses into existence, more than just the growth of environmental awareness has been noted. Not just the degree of information about critical ecological changes is increasing, but also a qualitative change is taking place. This becomes evident in a heightened awareness of environmental problems and the expansion of eco-friendly behavior and actions. In a recent survey about environmental awareness and behavior 64 % of the interviewed Germans claimed to have an excellent or good ecological knowledge (Schluchter, 1995). The wish of a considerable number of citizens for an ecological reform of the production- and consumption-processes is undeniable. On the other hand, though, an evidently immense need of security manifests in the orientation towards stability. Many people have observed and complained about a missing social concept, which can integrate changes with stability. This seemingly paradoxical atmosphere again doesn't meet with sufficient response by the political system. A social campaign to secure and improve environmental conditions, supported by large sections of society, can therefore not develop. As a consequence the individualization of ecological behavior is the most striking feature in dealing with environmental problems today. Very few people see the point of rolling up one's sleeves in order to bring about social changes, that would lead to a future ensuring ecological situation. At the moment the fear to endanger the already achieved, is greater than the hope for improvement.

In our survey of more than 1.000 representative respondents the answer to the question about their anxieties about problems may be broken down as follows: 7% 'were not anxious since solutions would be found to the problems'; 6% 'were not anxious because it was all useless anyway'; 72% 'were often anxious about the shape of things to come'; and 15% 'were very anxious about present and future problems'.

Considerable efforts for conservation of the environment are being made in some particular areas, like for example in the area of depositing refuse and sewage (consumptive waste avoidance and domestic sifting of waste). Yet, because of its individualization it doesn't seem to be as spectacular as the campaigns of the citizens' initiatives.

The environmental conservation is of great importance in the eye of the public. One could say, that the issue of ecological problems in industrial societies, has gained as much significance as the social question, concerning the distribution of the produced social wealth. This attitude indicates a new sense of direction, which interprets modernization ecologically and strives to be realized as a structural change in the process of industrial development.

The diagram below shows the degree of gravity with which specific problems are assessed (on a scale of 7 values).

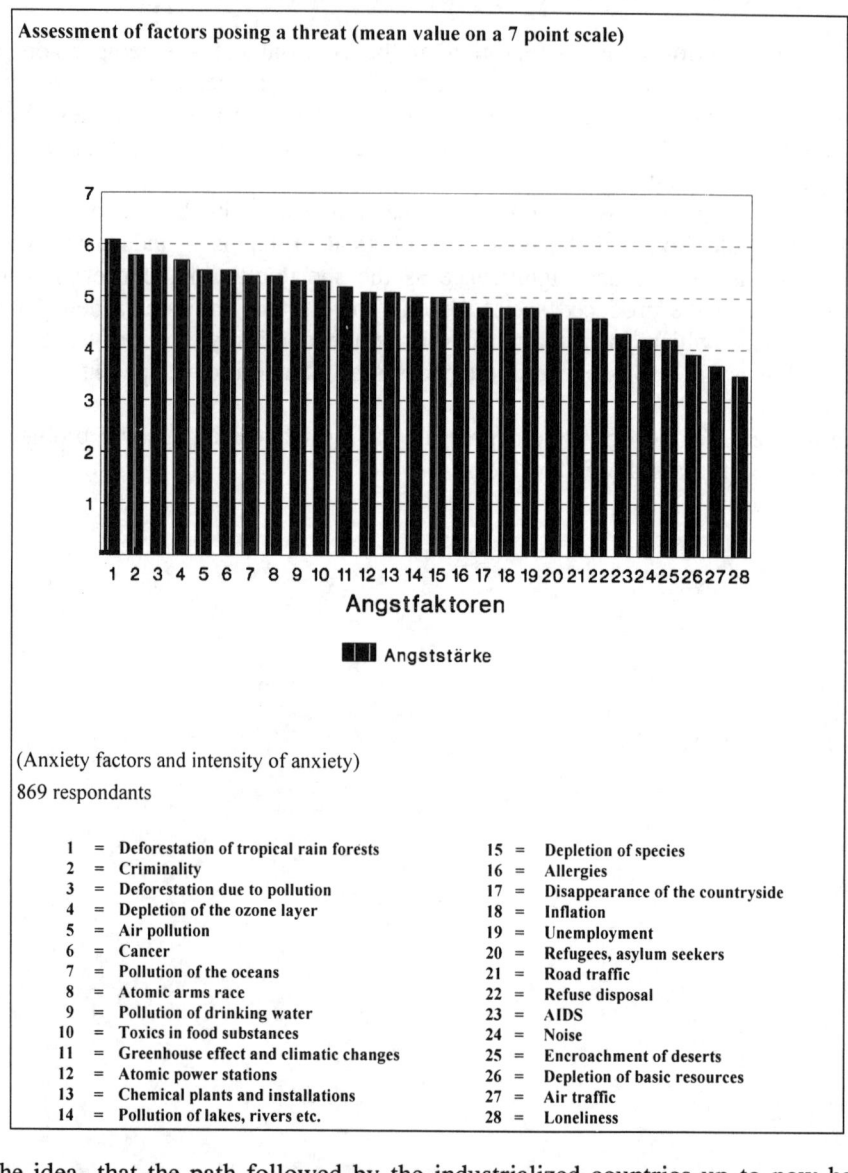

Assessment of factors posing a threat (mean value on a 7 point scale)

Angstfaktoren

■ Angststärke

(Anxiety factors and intensity of anxiety)
869 respondants

1	= Deforestation of tropical rain forests	15	=	Depletion of species
2	= Criminality	16	=	Allergies
3	= Deforestation due to pollution	17	=	Disappearance of the countryside
4	= Depletion of the ozone layer	18	=	Inflation
5	= Air pollution	19	=	Unemployment
6	= Cancer	20	=	Refugees, asylum seekers
7	= Pollution of the oceans	21	=	Road traffic
8	= Atomic arms race	22	=	Refuse disposal
9	= Pollution of drinking water	23	=	AIDS
10	= Toxics in food substances	24	=	Noise
11	= Greenhouse effect and climatic changes	25	=	Encroachment of deserts
12	= Atomic power stations	26	=	Depletion of basic resources
13	= Chemical plants and installations	27	=	Air traffic
14	= Pollution of lakes, rivers etc.	28	=	Loneliness

The idea, that the path followed by the industrialized countries up to now has its objective limitations, is widely spread in Germany. There are diverging opinions though, where these limits are. Many well-known experts belief the anthropogenic created effects of nature-manipulation and environmental structuring to be partly irreversible and partly manageable only by great social expenditure. Others are stating the opposite by referring to the technical progress and its potential to resolve the ecological problems. This creates mixed feelings of insecurity and a sense of living under constant threat, that gets stronger as the published opinions and statements polarize. Many people feel insecure and are afraid of what the future might bring. A widely spread loss of direction for individual and social endeavors, gives some the excuse to do nothing and others the chance to put the blame on somebody else. Others again regard immediate action as imperative. Many feel helpless and at the mercy of a process, which as individuals they can barely change

In some levels of society, particularly in the well-educated, materially better off middle classes, the desire for a reorientation of the technical progress is growing. These voices are getting louder, as the assessment of the foreseeable future is becoming more grave. Their incentive is not unconditional nature conservation, but rather the aspiration, to link economy and ecology in such a way, that it launches a future-oriented, ecologically responsible, and lasting development of society. Accordingly the economy shall be ecologically restructured, while simultaneously maintaining the present quality of life and providing the necessary security. The discussions about a differentiation, qualification or even revision of the existing growth concept, mark the beginning of a public debate about changing life styles. It is propelled by the discrepancy between post-material values, which hold more or less strong beliefs, and materialistic realities, in which we live. A renewal could secure the quality of life as well as contributing to the reorientation of the German growth ideal.

A solution for the global dilemma of dwindling resources and increasing damages to the environment, caused by traditional industrial progress strategies, is being looked for. This dilemma is well-known and becoming a topic of social debates and movements. It is definitely not anymore a reserved topic for scientific conferences, to discuss the destruction of the environment in association with a predictable scarcity of resources and increase in world population or to talk about possible mass migrations.

Many people try to bring the conditions of their environment under control by orientating towards easily comprehensible structures. Their efforts are to gain more autonomy over the protection and shaping of the individual conditions of their lives. This requires knowledge, although it does not need to be scientific. It is more a question of 'layman's logic', based on common knowledge and every day wisdom of the people. Often the 'expert's knowledge', based on scientific findings opposes the layman's logic and devalues it. Reversely the expert's knowledge is often being looked upon with mistrust. Consequently only that, which fits into a preconceived idea, is being accepted. According to the underlying interests the layman's logic and the expert's knowledge are either differentiated or joined together. A relative arbitrariness is the effect, which arranges itself according to results, that are deemed necessary. Hence the debates about possible prospects of development depend on opportune points of view, that in most cases are not even acknowledged, may be because of undeveloped particular social interests or because of limited problem awareness. A communication dilemma about the interests in question is characteristic for this situation. With increasing complexity of causes and effects this communication dilemma between experts and laymen gets even worse and the citizen ends up in an action dilemma. For the question now is, which kind of action or which kind of behavior is right or wrong, opportune or inadequate.

In Germany an ecological consciousness is developing as a consecutive process of the crisis (caused by inadequate behavior towards nature), that in its social constructed perception appears more and more as disastrous. However it doesn't develop without gaps or contradictions, found on all individual and political levels of action. They become especially obvious at the point where the politics of Germany and the European Union meet, that is, where the Federal Republic claims to be the first to do something and is really accepted in this role. An example is Germany's demand for emission-reduction by equipping cars with catalytic converters, when at the same time there is no speed-limit on German freeways. The slogan 'free speed for free people' ('freie Fahrt für freie Bürger') reflects a cultural phenomenon. In contradiction to this cause, the political approach is only a technological one. Another important point concerns the collapse of the East-German state and the associated improvement of the environment by shutting down the highly emitting industrial plants. The GDR had the world-wide strictest environmental laws, but they were hardly ever followed. It was mostly a purely ideological declaration, whose reality got refuted all the time in the eyes of the citizens. It generated such mistrust, disappointment and pessimism in the population, that it overshadowed even recognized social achievements and one can say, that the collapse of the GDR was not only caused economically and socially, but also by the ecological disaster, by which the East-German citizens saw themselves surrounded. In spite of the importance of resolving the environmental issues (which was recognized after the change) and the citizens' desire for a development of an affluent and promising society, the conventional social and economic problems (which in West-Germany seemed to be resolved or at least mostly settled) won priority. Serious social problems, like mass unemployment and the challenges of a recuperating development characterize the public consciousness of today more than the ecological challenges, which need to be resolved however, in order to make a future-oriented, promising process of development possible. After the change over in 1989 huge capital transfers took place from West-Germany to East-Germany. They stimulated mostly the sector of construction and private consumption. Short term satisfaction of needs of the citizens seemed to be a universal remedy for bringing both societies together. In the end, however, only those economical sectors experienced a considerable boom, that were competently occupied by West-Germany (e.g. the car-industry and the construction-machine sector). The opportunity for an ecologically oriented reconstruction was missed, that most probably would have secured more jobs than were cut. Also the opportunity for the reconstruction of industrial potentials, oriented towards the development and use of ecological know-how and technology (it could haven been tested and practiced simultaneously on East-German territory on a large scale), has been missed. Characteristically for the different levels of environmental awareness in East- and West-Germany was their respective reaction to the accident of the nuclear powerplant Tschernobyl. In West-Germany this accident met a highly sensitized awareness of the risks involved, developed during the conflicts about nuclear powerplants. Less than half of the population approved of civil use of atomic energy and after the incident the number went down to about 20 %. In East-Germany however, atomic energy generation was regarded as a great cultural and social improvement and only after Tschernobyl an awareness-process was triggered, comparable to the one that had started 20 years earlier in West-Germany. There is a considerable time difference in the development of environmental awareness and this partly explains the difference in newly formed life-styles in East- and West-Germany.

Environmental awareness in Germany expresses itself also in individual environmental behavior. According to our representative survey (Schluchter 1995) 33 % of people interviewed apply energy saving measures; 31 % claim to have eco-friendly mobility habits; 55 % opt for eco-friendly consumption-decisions, and 10 % engage themselves in a citizens' initiative or environmental group. One can observe the trend of individual environmental behavior turning into political action. It becomes apparent in the increasing political influence of the green party and the increasing demand upon the political system for creating an adequate framework for general eco-friendly behavior. This demand is based on the realization, that individual actions have a limited radius and are of little consequence. There is a section of society with above average awareness of ecological issues, that demands a future-oriented development and a certain standardization of eco-friendly behavior under the aspect of transformation of individual responsibilities into social ones. The political system cannot escape from these demands and there are already some responses, though they are not yet up to that standard, as many citizens would like.

The diagram below gives a break-down of answers to the question how competent do respondents rate groups or institutions to provide solutions to problems.

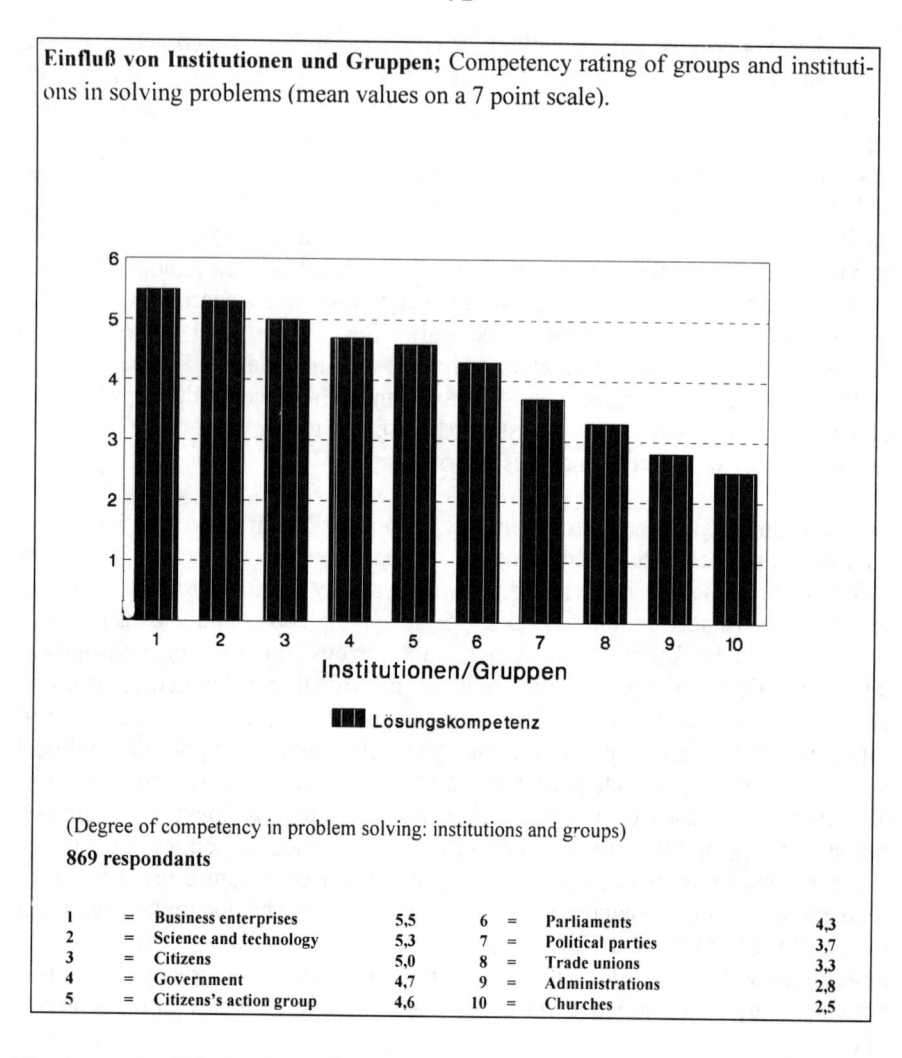

Einfluß von Institutionen und Gruppen; Competency rating of groups and institutions in solving problems (mean values on a 7 point scale).

(Degree of competency in problem solving: institutions and groups)
869 respondants

1	=	Business enterprises	5,5	6	=	Parliaments	4,3
2	=	Science and technology	5,3	7	=	Political parties	3,7
3	=	Citizens	5,0	8	=	Trade unions	3,3
4	=	Government	4,7	9	=	Administrations	2,8
5	=	Citizens's action group	4,6	10	=	Churches	2,5

The loss of solidarity in society as a result of increasing individualization and disassociation contributes significantly to loosing the sense of responsibility for systemic correlations and strengthens egocentricity. 'Everybody cares just about himself', is a complaint one often hears, referring to the failure of collective, social solutions of problems. Missing personal communication between individuals reinforces this phenomenon and obstructs an encounter between the different views, that could help clarify the issues. It hinders the formation of social standards and norms for the realization of collective, social interests. What remains, is just the individual interest, limited by norms, that aren't defined by the individual itself. This constitutes a serious deficiency for the autonomous evolvement of responsibility for ecological conditions.

One can propose the hypothesis, that there is considerably more willingness for eco-friendly behavior, than is factually being practiced. The question arises, what is the hindrance? One could look at this phenomenon from the angle of a model of rationality or a model of anxiety. One variation could be: if only a few do something, the individual contribution becomes irrational, as it requires much and turns out little. Another variation is based on the assumption, that the species man will disappear from this planet before long and that no act can alter this fact. Often one hears, that the citizens expect from the political system guiding principles, generally binding for all individual and social endeavors for environmental conservation. They demand clear and reconstructible targets with visible and perceptible effects. Numerous economists, who are opposed towards the economic council, emphasize the fact, that environmental conservationand protection of the resources is not needed solely for its own sake, but that it gives also a perspective for the future economy.

There are indications for a diverse environmental awareness in different professions. Senior professions, especially those in the service sector - which is developing in Germany into the most important economical sector - show greater environmental awareness, than traditional professions in the production sector (here the combination of existential security needs, expressed in the anxiety about loosing one's job and a pessimistic attitude , as it is publicized by the media, shows its effects). Executives in the service sector discover frequently, that the traditional industrial rationality, previously a powerful motor for future developments, has turned into a stumbling block. And here they see their chance. Because for them modernization is a factor for social promotion and affluence.

The modernization tendency in Germany however doesn't end with these expectations. It is also concerned about casting off those old attitudes, that are leading to ecological damages, requiring more expenditure for their restoration than they contribute to the well-being or satisfaction of needs. One can say, a process of weighting the economic profits has begun. Its result is the demand for a say in the distribution and utilization of the societal resources and the unwillingness of many citizens to accept the annual budgets of the government without opposition.

Elections with a high number of non-voters seem to signal at first glance disinterest in political decisions (in Germany one talks about the political frustration of at least 30 % of eligible voters), but the active participation in communal decisions is increasing and quite frequently energetically claimed. It is noticeable, that the political authorities tend to disapprove of these claims, and by doing so produce negative psychosocial effects in the citizens, who see their efforts cut off from the political system and start feeling frustrated. The result is either an escapism into their private sphere or the attempt to win comrades-in-arms. Accordingly there is a interrelation between private sphere and social movements, that maps out the social tendencies for change. And this is the real modernization shift in German society.

This modernization shift explains the not anticipated powerful protest of the Germans against Shell's planned submersion of its oil-platform. Differently from other countries - where a similar protest attitude did not exist and where on the contrary lack of understanding about the strange behavior of the Germans was being conveyed - many took the opportunity to express their strive for autonomy and modernized attitude. Yet the pollution of German rivers is many times bigger than the one, the submersion of the oil-platform would have caused, and this provokes only partially an open protest. It would not occur to anybody, to boycott the responsible industrial enterprises, as it was done in the case of Shell. Prime movers for the protests against Shell, where their high-handed and arbitrary manner of action; the full coverage of the media, spawned by the campaign of Greenpeace; and Shell's obvious moneysaving motive. These incidents were taken in with a sensitized environmental awareness and related to the outlook of modernism, that reveals such measures as old-fashioned and irrational. To this factors one can add the opportunity to express a protest collectively and thus overcome the limitations of individual protest. The aim of the protest was obvious; the boycott of Shell required little effort; the collective 'we-feeling' was supported by statements of high-ranking politicians, who joined the protest later on a high level and thus sanctioned the individual behavior. This boycott demonstrated the modern form of changed life-styles.

A similar development is taking place in the protest against the French atom-bomb tests. Obviously nobody thought about the modernization of life-styles and its unfolding forms of expression when these tests were being planned. But their intention might be to test, how far the restoration can go and push its project through against the modernization. It will be quite interesting to see, if this episode will initialize also in France - with the support of internationalized protest - modern, ecologically oriented life-styles or if it will strengthen the nationalistic Right.

4. CONCLUSIONS

In the Federal Republic of Germany environmental awareness and behavior has evolved within dynamics of concern, worry, scandal, problematic issues and information. They are the result of social conflicts and debates, which induced a change of model for social development. This change steers away from the development-model of increasing material affluence in society at all costs, achievable only through constant quantitative production-growth of industrial goods. Alternately an ecological paradigm of a new model of affluence with productive and lasting accomplishments is unfolding. This is noticeable not only in the context of changing individual attitudes and models of behavior, but also in the change of direction towards an ecologically oriented industrial production. The strategy of clinging to the conventional logic of industrial production, with the use of technologies, characterized by process accompanying emissions, that afterwards are being reduced, filtered or neutralized, seems to have reached in many cases its dead end. Further progression requires disproportionate economical and technical expenditure, that doesn't pay off profitably. In perspective however is a strategy of ecological issues, integrated into a new logic of industrial production, that shows by far more favorable relations between expenditure and profit, with simultaneously less emissions and better use of resources and energies. This new paradigm would also be exemplary for development-strategies and -processes of the Third World. The positive effect and benefit for the Federal Republic of Germany lies in the restoration of economic income under conditions of restricted use of resources and the improvement of the environmental quality. Both upgrade the standard of life and thus create a stronger identification of the citizens with their state. Also a better understanding of the necessity of ecologically oriented reforms is starting to show. It is because of these understandings, that the conventional attitudes, concerning national economy models as well as the traditional expenditure-profit attitudes are on the decline. The political development in Germany however is lagging behind. The political system does not develop suitable structural and economical prospects and models in a satisfactory measure, nor does it produce a link between individual behavior and action-stimuli and the goal of a modern ecological and social action paradigm. It rather limits itself towards merely symbolical environmental politics, based on noncommittal declarations and intentions. A good example is the German declaration about reducing 25 % of its carbon dioxide emissions, given by the German delegation in Rio. An important characteristic of this symbolic environmental politics is also the government's demand of the Third World countries to shape and organize their developing processes in an ecological way, which in view of the growing world-population is just a shift of the guilt question. The government of the Federal Republic just casts off its own responsibility and by doing so maneuvers itself into a contrary situation to large sections of its population, who are ready for a reorientation. The Federal Republic not only possesses the necessary economical resources and scientific-technological potential to make the forerunner for others as far as the conservation of the environment is concerned, but it also holds the possibility to develop into an ecological model for the western industrialized countries. The spreading environmental awareness of the population as well as the growing sensitization in the economic system point to this fact. Environmental awareness of the citizens becomes an important location factor. The political change however takes its time, although conceptual elements of a new era are expanding inside the political system.

REFERENCES

Amery, C. (1978). *Natur als Politik - Die ökologische Chance des Menschen.* Hamburg: Reinbek b.
Aurand, K., B. Hazard and F. Tretter. (Hg.). (1993). *Umweltbelastungen und Ängste - Erkennen, Bewerten, Vermeiden.* Opladen.
Beck, U. (1986). *Risikogesellschaft - Auf dem Weg in eine andere Moderne.* Ffm.
Beck, U. (1988). *Gegengifte - Die organisierte Unverantwortlichkeit.* Ffm.
Beck, U. (Hg.). (1991). *Politik in der Risikogesellschaft.* Ffm.
Bogun, R., M. Osterland and G. Warsewa. (1990). *Was ist überhaupt noch sicher auf der Welt? - Arbeit und Umwelt im Risikobewusstsein von Industriearbeitern.* Berlin.
Böhme, G. (1991). 'Die Frage nach einem neuen Naturverständnis'. *Politische Ökologie* 24, Nov./Dez.
Dahl, J. (1991). 'Verteidigung des Federgeistchens - Über Ökologie und Ökologie hinaus'. *Politische Ökologie* 24, Nov./Dez.
Deutscher Bundestag (Hg.). (1988). *Schutz der Erdatmosphäre - Eine internationale Herausforderung, Zwischenbericht der Enquete-Kommission 'Vorsorge zum Schutz der Erdatmosphäre'.* Bonn.
Deutscher Bundestag (Hg.). (1993). 'WBGU - Wissenschaftlicher Beirat der Bundesregierung "Globale Umweltveränderungen", Welt im Wandel - Grundstruktur globaler Mensch-Umwelt-Beziehungen'. *Bericht der Bundesregierung, Jahresgutachten 1993.* Bonn.
Diekmann, A., Franzen, A. (Hg.). (1995). *Kooperatives Umwelthandeln - Modelle, Erfahrungen, Maßnahmen.* Chur/Zürich.
Dietz, F.J., U.E. Simonis and J. van der Straaten. (Hg.). (1992). *Sustainability and Environmental Policy.* Berlin.
Doran, C. F., Hinz, O. Manfred and P.C. Mayer-Tasch. (1974). *Umweltschutz - Politik des peripheren Eingriffs - Eine Einführung in die politische Ökologie.* Darmstadt/Neuwied.
Dreitzel, H.P. and H. Stenger (Hg.). (1990). *Ungewollte Selbstzerstörung, Reflexionen über den Umgang mit katastrophalen Entwicklungen.* Ffm./N.Y.
Enzensberger, H.M. (1973). 'Zur Kritik der politischen Ökologie'. *Kursbuch* 33, S.1ff.
Fietkau, H-J. (1984). *Bedingungen ökologischen Handelns.* Weinheim.
Fietkau, H-J. and H. Kessel. (Hg.). (1981). *Umweltlernen.* Königstein/Ts.
Glaeser, B. (1989). *Umweltpolitik zwischen Reparatur und Vorbeugung.* Opladen.
Hartkopf, G. and E. Bohne. (1983). *Umweltpolitik.* 2 Bde., Opladen.
Heine, H. and R. Mautz. (1989). *Industriearbeiter contra Umweltschutz?* Ffm.
Huber, J. (1989). *Technikbilder - Weltanschauliche Weichenstellungen der Technologie- und Umweltpolitik.* Opladen.
Huber, J. (1993). 'Ökologische Modernisierung - Bedingungen des Umwelt-

handelns in den neuen und alten Bundesländern'. *Kölner Zeitschrift für Soziologie und Sozialpädagogik* 46, Heft 2/1993.

Jänicke, M. (Hg.). (1978). *Umweltpolitik.* Opladen.

Jarre, J. (1993). 'Belohnungen: ein noch zu entdeckendes Instrument zur umweltbezogenen Verhaltensbeeinflussung privater Haushalte'. *Zeitschrift für angewandte Umweltforschung* 6/1993.

Jungermann, H., B. Rohrmann and P.M. Wiedemann. (Hg.). (1991). *Risikokontroversen - Konzepte, Konflikte, Kommunikation.* Berlin.

Kessel, H. and W. Tischler. (1984). *Umweltbewußtsein.* Berlin.

Maier-Rigaud, G. (1988). *Umweltpolitik in der offenen Gesellschaft.* Opladen.

Luhmann, N. (1986). *Ökologische Kommunikation.* Opladen.

Luhmann, N. (1991). *Soziologie des Risikos.* Berlin, N.Y.

Metzner, A. (1993). *Probleme sozio-ökologischer Systemtheorie - Natur und Gesellschaft in der Soziologie Luhmanns.* Opladen.

Metzner, A. (1994). 'Offenheit und Geschlossenheit in der Ökologie der Gesellschaft'. In: F. Beckenbach, H. Diefenbacher, *Zwischen Entropie und Selbstorganisation - Perspektiven einer ökologischen Ökonomie.* Marburg: S.349-391.

Müller, E. (1986). *Innenwelt der Umweltpolitik - (Ohn)macht durch Organisation?* Opladen.

Naujoks, F. (1991). *Ökologische Erneuerung der ehemaligen DDR - Begrenzungsfaktor oder Impulsgeber für eine gesamtdeutsche Entwicklung?* Bonn.

Pongratz, H. (1992). *Die Bauern und der ökologische Diskurs - Befunde und Thesen zum Umweltbewußtsein in der bundesdeutschen Landwirtschaft.* München, Wien.

Prittwitz, V. von. (Hg.). (1990). *Das Katastrophenparadox - Elemente einer Theorie der Umweltpolitik.* Opladen.

Prittwitz, V. von. (1993). *Umweltpolitik als Modernisierungsprozeß - Politikwissenschaftliche Umweltforschung und -lehre in der Bundesrepublik.* Opladen.

Rat von Sachverständigen für Umweltfragen. (1994). *Umweltgutachten 1994 - Für eine dauerhaft-umweltgerechte Entwicklung.* Bonn.

Reichert, D. and W. Zierhofer. (1993). *Umwelt zur Sprache bringen - Über umweltverantwortliches Handeln und den Umgang mit Unsicherheit.* Opladen.

Ruff, F. (1990). *Ökologische Krise und Risikobewußtsein - Zur psychischen Verarbeitung von Umweltbelastungen.* Wiesbaden.

Schluchter, W. (1991). *Die psychosozialen Kosten der Umweltverschmutzung, Umweltbundesamt.* Texte 24/91. Berlin.

Schluchter, W. (1995). *Analyse der Bedingungen für die Transformation von Umweltbewußtsein in umweltschonendes Verhalten, Umweltbundesamt.* Berlin.

Sieferle, R-P. (1984). *Fortschrittsfeinde? Opposition gegen Technik und Industrie von der Romantik bis zur Gegenwart.* München.

Simonis, U.E. (Hg.). (1988). *Präventive Umweltpolitik.* Ffm, N.Y.

Simonis, U.E. (Hg.). (1988). *Lernen von der Umwelt - Lernen für die Umwelt - Theoretische Herausforderungen und praktische Probleme einer qualitativen Umweltpolitik.* Berlin.
Trepl, L. (1983). 'Ökologie - eine grüne Leitwissenschaft?'. *Kursbuch* 74:6-27.
Weizsäcker, E.U. v. (1989). *Erdpolitik - Ökologische Realpolitik an der Schwelle zum Jahrhundert der Umwelt.* Darmstadt.
Wey, K-G. (1982). *Umweltpolitik in Deutschland, Kurze Geschichte des Umweltschutzes in Deutschland seit 1900.* Opladen.
Wörndl, B. (1992). *Die Kernkraftdebatte - Eine Analyse von Risikokonflikten und sozialem Wandel.* Wiesbaden.

6. The Evolution of the Soviet/Russian Ecological Movement: Political Trends

Oleg N.Yanitsky
Irene A.Khalyi

Abstract

The ecological movement in the USSR/Russia exists for 35 years and remains one of the most stable social actors in the country's political arena. This chapter describes the sources and the forms of political activity of various groups belonging to the movement, and analyses the stages of this activity using the concept of 'political opportunity structure'. Special attention is paid to the relationships between the conservationist and the national patriotic wings of the movement. The tendency toward its bureaucratization is shown, proving that the movement is becoming less radical and more reformist.

1. INTRODUCTION

The ecological movement is one of the most stable social actors in the Russian political arena. For as much as 35 years now, it has been developing, differentiating, and using the growing number of various tactics. That is why a close study of its evolution as well as its political behavior at various stages seems timely. Such analysis is impossible without the simultaneous consideration of the dynamics of the social and political contexts and the changes in the 'political opportunity structure'. The structure itself consists of five major blocks: 'degree of openness/closure of the polity; stability/instability of political alignments; presence/absence of allies and support groups; divisions within the elite or its tolerance for protest; policy-making capacity of the government' (Tarrow, 1988:429). Taking this very concept as the point of departure, it is important to note that the behavior of the collective social subject (such as, e. g., the ecological movement) is far from being completely determined by the above structure. Due to its own actions the spectrum of its political capacities may be either widened or narrowed. Besides, the structure is not limited by legal frames and existing institutions. It changes in the course of direct actions, infiltration, and other forms of unconventional political behavior.

2. THE EMERGENCE OF THE ECOLOGICAL MOVEMENT

The ecological movement emerged in the 1960s, the period of the first 'perestroika' in the USSR. Mass repressions had already become a thing of the past, the state economy seemed thriving, political and social life became more democratic. The CPSU appealed to the Soviet citizen to exhibit more initiative, taking part in the 'great constructions of communism' and economic competition with the USA. The second wave of industrialization started: the country's chemical industry was established, new nuclear power plants and hydro-power stations were constructed, the vast spaces of virgin steppes and taiga were rapidly transformed into agricultural and industrial areas. Yet the disastrous consequences of this second major attack of the Soviet state machine upon Nature (the first one having been the industrialization and collectivization of the 1930s) soon became visible. Starting from 1958, the public began regular protests against the pollution of the Baikal, the largest freshwater lake in the world. Those protests were initiated first and foremost by the intelligentsia. Collective letters of concern and appeals to the press, resolutions of public meetings and conferences on the Baikal issue, as well as other forms of protest sanctioned by the rulling regime, proved to be so powerful and steadfast as to make the Central Committee of the CPSU and the USSR Government issue decrees concerning the preservation and rational use of the Baikal natural resources in 1969, 1971, 1977, and 1987, respectively (Lapin, 1987). In our view, the political success of the Baikal movement may be explained by several reasons: the existence of the cells of civil society in the pores of the totalitarian state, those, in particular, that emerged in Siberia, in the North and the Far East, where thousands of scientists, engineers and other specialists were sent to exile in Stalin's times; the realization of state programs of the development of science in Siberia, in particular, the establishment of the Siberian branch of the USSR Academy of Sciences and its own sections in the cities of Irkutsk and Krasnoyarsk; the rapid growth of environmental concern among the local intelligentsia which succeeded in creating a well-developed network of non-governmental organizations in Siberia, and, in general, enjoyed more freedom of collective actions in comparison with Moscow intellectuals. The 1960s-1980s also witnessed the emergence of the broad social basis of the ecological movement: the civil initiatives. What is here meant by this term is the groups of residents acting in their neighborhoods, whose goals included making their immediate environments clean and safe.

The communist ideology and Soviet urban policy played direct role in shaping those cells. The CPSU called for more social activity of the citizen in their immediate environments. The grandiose house-building campaign, launched in the 1960s all over the USSR, made it necessary to introduce certain elements of self-organization into the life of newly-emerged neighborhoods. The CPSU cells were created there in great numbers. A natural process of 'city ageing' brought along the increase and the concentration of pensioners in those neighborhoods. At the same time, families with small children were rapidly growing in number especially in newly-built towns. The communist regime widely supported the patronage of local enterprises over 'their own' neighborhoods, schools, clubs, etc. As a result of that and other reasons, the social life of cities and towns could no longer be totally controlled by the CPSU cells. Moreover, the municipal services, unable to cope with their duties, began gradually to transfer their responsibilities and some of their functions onto the housekeeping offices established to serve houses. It was at that very time that the CPSU ordered to turn 'election points' (that is, local cells for supervising elections to the Soviet power bodies) into acting centres for political and cultural work with the local population. What emerged, then, was an odd economic-cultural-political hybrid. Gradually, however, its official political activity was fading away, while social and cultural work, on the contrary was developing, even gaining certain new sides. It was in this milieu that the figures of local leaders began to shape. They started to put forward their own organizational and cultural initiatives, and then ecological and political ones as well (Yanitsky, 1991).

Students' Nature Protection Corps turned out to be the third major component of the ecological movement in the USSR. The first of them were established at the biological departments of the Tartu and Moscow Universities in the early 1960s. The professional and social life of those departments became a sort of 'feeding ground': they were the permanent source for the recruitment of new members of the Corps, provided them with all necessary resources, and carried out scientific guidance over their activities. Initially Students' Nature Protection Corps were controlled by the departmental Young Communist League organizations (YCL). However, the more sophisticated became the tasks these Corps were working on (such as the establishment of nature reserves, seasonal scientific observations of natural phenomena, ecological propaganda, campaigns against poaching), the more difficult it became for the YCL to exercise real control from its political 'top', until it finally turned into a mere formality. On the other hand, the Corps, having gained strength and established ties with one another, began persistent struggle for their independence, both organizational and ideological. Such a struggle for their independent activities was also facilitated by the fact that the Corps could perform their duties under the auspices of more formal public organizations, namely the All-Russian Society for Nature Protection, various public inspections, hunting clubs, etc. By the beginning of the 1980s, the network of the above Corps, though still remaining an elite university structure, became a powerful social force. Its Charter proclaimed that 'the work in the Corps is a form of social-political practice', and the Corps itself, as a collective member of the All-Russian Society for Nature Protection, 'organizes its activities on the basis of self-government and self-education' (Mukhachev and Zabelin, 1992:123). Like the mentioned civil initiatives, the Corps' movement was not particularly a protest kind. On the contrary, its main goal was to assist state bodies and official public organizations in nature protection. But this very participation let the Corps' members gain enough knowledge about the real scale of environmental destruction and anti-ecological practice of the state, and stimulated the accumulation of the potential for social protest which was eventually realized during the perestroika times.

4. ELECTIONS TO THE ALL-UNION AND REPUBLICAN PARLIAMENTS, 1989-1990

In 1989-90, the emergence of ecological groups and movements continued: during that period, for instance, the All-Union voluntary society 'Green Movement' was established , and green parties as well as the like movements sprang up in the Ukraine, Byelorussia, aand the Baltics. In February, 1989, for the first time in the history of the USSR, more than 300 thousand residents in 100 cities all over the country simultaneously took part in the mass antigovernment protest action against further construction of the Volga-Chograi Canal, once a constituent part of the mentioned project of transferring part of the Northern rivers' flow to the Volga basin. The action was successful: the USSR government ordered to close down the construction. Yet, the main political event of those years was arguably the first democratic elections to the USSR People's Deputies' Congress (the All-Union Parliament). This event considerably transformed and diversified the activities of the ecological groups. To begin with, their political activities as such (taking part in the pre-election meetings and conferences, nominating 'green' candidates, helping them prepare their pre-election programs, etc.) became more and more detached from their social and cultural work. A certain pattern of collective actions emerged: protest action — establishment of local (regional) movement — participation in pre-election campaigns. Secondly, these protest actions acquired an evidently *instrumental* character, thereby removing the function of the movement's *social identification* to the background. As a matter of fact, it was the beginning of the movement's split into political and non-political fractions, or into *leaders*-politicians and *activists*-experts. Furthermore, the deeper the movement plunged into political process, the deeper its inter-group differentiation became. Last but not least, the pre-election campaigns revealed a problem which was very soon to become the key political issue for the ecological movement throughout the USSR. By this issue we mean the relationships between the ecological and the national-patriotic movements. In the 1989 elections to the All-Union Parliament the national-patriotic forces suffered a defeat. However, in the 1990 elections to the parliaments of the Republics and local Soviets these forces not only managed to come to power under the slogans of saving national nature and fighting the colonial policy of the Center, but also made use of the electorate of green organizations, thus depriving them of their social basis (Fomichov, 1993; Boreiko and Listopad, 1994). In a number of cases, the former ecologists joined the nationalist movements, a striking example of which transformation happened in Moldavia (Mikhailov, 1992). Concerning the 1989 elections, our estimates show that about as much as 300 ecologically oriented deputies were elected to the All-Union Parliament, which constituted 12% of its total membership. Forty of them were the acknowledged leaders of various ecological groups and movements. One would consider this as indicative of promising good prospects for establishing a powerful 'ecological core' (or a strong fraction) in the Parliament, especially given that ecological values kept on playing a consolidating role within that assembly. However, this did not happen. Ecological issues were 'drowned' in general political debates. Only a year later the exceptional persistence of a few deputies-ecologists moved the USSR Supreme Soviet to adopt the decree 'On urgent measures to heal the environment'.

5. 1990-1992: COOPERATION WITH THE AUTHORITIES; THE BEGINNING OF THE MOVEMENT'S DEEP TRANSFORMATIONS

The main event of this period — the collapse of the USSR and the emergence of new sovereign states on its territory — hit a serious blow on the ecological movement. The attention of both the public and all political forces was completely absorbed by a new political and economic situation, political priorities shifted to the issues of state-building and developing the inter-state ties. The socio-economic context in all former republics of the USSR was becoming less and less favorable for the political activity of Greens. Living standards were rapidly declining, and the population found itself deprived of many of those social guarantees which had long been habitual. Unemployment increased, most dramatically among the youth and women. Violent armed conflicts began in many Central Asian and Trans-Caucasian republics. Another negative factor that emerged during this period was the revision by these new states, including Russia, of their ecological policy. The majority of polluting enterprises, including nuclear power plants, recently closed down or temporarily shut under the pressure of Greens, were put into operation once again. Moreover, the decisions were often made by those persons who had become government officials or parliament members owing to the support they had received from the ecological movement. This unabashedly careerist behavior was a profound psychological shock for the movement's rank-and-file activists. Besides, certain leaders of the ecological movement grew more and more loyal to the national-patriotic ideology. The same processes took place in Latvia, Estonia, Lithuania and Moldavia. On the whole, the process began which may be called the *frontal retreat* of the new national politicians from the ecological programs they had declared. That meant a considerable narrowing of Greens' political opportunity structure and their loss of habitual ways of mobilizing resources.

So what were the political tactics of Greens during this period? Before the collapse of the USSR and then in Russia of 1991-92, their major organizations (namely, the Socio-Ecological Union, the Ecological Union and the 'Green Movement'), counted on the tactics of cooperation with legislative and executive powers in the Center as well as the provinces. It is very important to stress here that, as a rule, the *initiative* of working out various normative acts that concerned ecological issues *came from those power structures* themselves. Some ecoactivists, while remaining the movement's members, simply took up jobs at the Presidential apparatus, the Supreme Soviet of the Russian Federation; others became deputies of local Soviets or officials of the newly-established Ministry of Environment and for the last ten years the leading core of the movement, consisting of professional biologists, considered the making of the movement's political strategy as its own prerogative, and therefore strongly opposed any attempts of eco-sociologists and political scientists to take part in this task. The problem of the leaders' incompetence in the theoretical aspects of green politics has not been overcome until today: suffice it to say that, of the three newly-established organizations, the movement 'For Chemical Security' is headed by a chemist, the 'Russian Green Cross', by a mathematician, and the leader of the international movement 'Green Cross' is Mikhail Gorbachev whose works on the theory of green politics are hardly known to anyone.

The second of the above-mentioned reasons, the movement's differentiation, is also rather serious. As we have shown earlier, each of the seven branches constituting the Russian ecological movement (conservationists, alternativists, traditionalists, civil initiatives, eco-politicians, eco-patriots, and eco-technocrats) has its own structure of political opportunities and specific relationships with the state, the public, and the institutions of civil society (Khalyi, 1994; Yanitsky, 1994). Now let us consider a more complicated question — namely, *how the movement's leaders envision the strategy of transition to sustainable development for the entire society*. For instance, let us take a look at how this strategy figures in the project prepared by the SEU Council. Even its very first statements sound disturbing: 'Development is an increase in the amount of happiness (health, security, and intercourse), not in the amount of things; (. . .) Development is sustainable if the children are happier than their parents; (. . .) Everyone knows what he needs to be happy better than his boss'. Why is the definition of sustainable development given here in moral and ethical terms? Where are the basic economic, demographical, and other criteria? And what indeed does the 'boss' have to do with all that? 'National wealth can grow only when the same needs are satisfied with the less expenditure of time, materials, and energy.' To be sure, frugality is a very important criterion. But where is the criterion of efficiency? And what are we to do with our physical needs? Should we eat less or save money at the expense of our health, because its preservation also demands the 'expenditure of time, materials, and energy'?

As we proceed further on, we encounter obvious contradictions. On the one hand, it is stated that 'Russia has all conditions for sustainable development by means of its own resources'. On the other, 'sustainable development of humanity on the Earth can only be attained by the joint efforts of all peoples'. For one thing, the project assumes the necessity of reducing the areas subject to economic use. In other words, the intensification of agriculture is meant as a condition for restoring the balance of the global ecosystem. For another thing, 'All citizens of Russia must be provided with plots drawn from agricultural lands, which would be large enough to procure them with self-grown food stuffs', etc. All in all, the project has lots of details, but lacks the main point: what type of society should be created, and how the gap between the critical present and the 'bright' ecological future is to be bridged? One (shocking) point in this regard can be found, however: 'A considerable part of work providing Russia's ecological security may be commissioned (and, consequently, financed from public funds) to the military forces and state security service' (SoES..., 1994). Incidentally, this standpoint has much in common with the view expressed by the chair of 'Russian Green Cross', that the nuclei of future civilization, which need to be preserved at whatever cost, are the enterprises of elite industries — first of all, of the military-industrial complex (Moiseev, 1994:156, 158).

7. CONCLUSIONS

The 1985-95 decade of reforms in the USSR/Russia witnessed constant changes in the political opportunity structure. At the initial stage of these reforms the openness of Soviet society grew, whereas today Russian society is becoming more and more closed. The period of democratic upsurge (1989-91) was marked by 'ecological solidarity' of Soviet society. But later the political positions won by Greens in the course of mass protest actions, were lost. National-patriots, on the contrary, widened their political opportunities. The overall political instability has been growing. The possibilities for institutionalization of the public's ecological demands are minimal. The entire political context becomes more and more hostile to the movement's goals and values. Within the movement, two wings become more and more manifest: nature conservationists whose leaders are somewhat reluctantly drawn into politics, and the political wing oriented toward struggle for power. The first wing is gradually removed from the political arena by the second one in which 'state-favoring technocrats' find common language with 'patriots'. However, both wings continue to view eco-politics mainly in terms of nature protection and preservation of national resources. Consequently, they are willing to develop an 'isolated' eco-politics, but unwilling to change the strategy of societal reforms as a whole. The more difficult and unstable the political situation becomes, the clearer becomes the orientation of both wings of the movement toward identification with the state as the main political force. The nature-conservationist wing has lost its mass social basis and, with it, its ability to mobilize the public for mass protest actions. Nature conservationists avoid cooperating with housing movements, movements for the defence of customers' rights, local self-government, and other akin organizations. 'KEDR' and other state-favoring ecological movements derive their constituency and infrastructure directly from the cells of state and municipal services. 'Patriots' appeal to the military and simultaneously to pensioners and other most deprived strata. The elites of all branches of the movement avoid direct confrontations with the state. Mass protest actions has been practically excluded from the movement's tactical arsenal. Radical actions of eco-anarchists and threats of mass riots coming from 'patriots' do not change the political situation as a whole. The government today has no articulate ecological policy. The President has set the goal of transition to sustainable development, but this goal is no priority for those who actually make politics in Russia. The ecological movement has not put forward its own alternative model of transition to sustainable development. In conclusion, it may be stated that (1) the movement has failed to use existing political opportunities, even those created by itself; (2) its prevalent orientation was toward correcting the 'mistakes' of the regime; and (3) it has been less inclined to political confrontations with the regime less than other new social movements. Most recently, two complementary tendencies have emerged: direct servicing of the state needs by ecological NGOs, and the organization by the state of 'its own' quasi-public green parties and movements, totally dependent on it.

REFERENCES

Boreiko, V. and O. Listopad. (1994). 'Zelenye na Ukraine. Vchera. Segodnia. Zavtra? *Tretii Put'* 36:4-9.
Downs, A. (1972). 'Up and Down with Ecology - the Issue Attention Cycle'. *Public Interest* 28:38-50.
Ekologicheskaya Bezopasnost' Rossii. (1994). Vypusk 1. Moskva: Yuridicheskaya Literatura.
Fomichov, S. (1993). 'Again to the Crisis Question'. *The Third Way* 4:3-6.
Fomichov, S. and L. Galkina. (1993). 'A Short History of the Party (Greens)'. *The Third Way* 4:7-9.
Khalyi, I. (1994). 'Environmental and Nationalist Movements: Allies or Adversaries?' *Paper presented at the session of RC N 24 'Environment and Society', World Congress of Sociology.* Bielefeld, Germany, July 17-24.
Lapin, B. (ed.). (1987). Slovo v Zashchitu Baikala. Materialy Diskussii. Irkutsk: Vostochno-Sibirskoe Knizhnoe Izdatel'stvo.
McAdam, D., J. McCarthy and M. Zald. (1988). 'Social Movements'. In: N.J. Smelser, (ed.) *Handbook of Sociology*. Newbury Park: Sage Publications, p. 695-737.
Mikhailov, V. (1992). 'Natsional'noe Dvizhenie: Moldavskii Variant'. *Polis* 4:85-93.
Moiseev, N. (1994). 'Prichiny Krusheniia i Rychagi Protsvetaniia'. *Rossiiskaya Provintsyia* 2:153-164.
Mukhachev, S. and S. Zabelin. (eds.). (1992). *30-let Dvizhenia. Neformal'noe Prirodoochrannoe Molodezhnoe Dvizhenie v SSSR. Fakty i Documenty. 1960-1992.* Kazan': Sotsial'no-Ekologicheskii Souz.
'SoES: Kontseptsia Ustoichivogo Razvitia Rossii'. (1995). *Zelenyi Mir* 5:5-6.
Sotsial'no-Ekologicheskii Souz: Istoria i Real'nost'. (1994). Moskva: Tsentr Koordinatsii SoES.
Tarrow, S. (1988). *Democracy and Disorder. Protest and Politics in Italy, 1965-1975.* Oxford: Clarendon Press
Yanitsky, O. (1991). 'Lefortovo, Moscow: Resolving Conflict Between Urban Planners and Residents'. In: T. Deelstra, O. Yanitsky (eds.), *Cities of Europe: the Public's Role in Shaping the Urban Environment.* Moscow: Mezhdunarodnye Otnoshenia Publishers, p. 356-371.
Yanitsky, O. (1993). 'Environmental Initiatives in Russia: East-West Comparisons'. In: A. Vari, P. Tamas (eds.), *Environment and Democratic Transition. Policy and Politics in Central and Eastern Europe.* Dordrecht: Kluwer Academic Publishers, p. 120-145.
Yanitsky, O. (1994). 'Ekologicheskaya Politika: Rol' Dvizhenia i Grazhdanskih Initsiativ'. *Sotsiologicheskie Issledovaniia* 10:10-20.

7. Support for Environmentalism During a Major Recession

Elim Papadakis

Abstract[1]

In the social sciences we are often confronted by limitations in the way we think about problems. For instance, in the area of environmental policy, if you are for economic development, you are often regarded as opposed to environmental protection. This way of thinking would lead us to expect a decline in support for environmental protection during an economic recession. Much of this is linked to the way in which we tend to adopt 'binary codes' (in this case 'environment' versus 'development') to simplify a complex environment (Luhmann, 1990). Data from several Australian Election surveys are used to explore several hypotheses. Paradoxically, the environment remains an issue of great concern to most people. An underlying reason for the reduced emphasis on the environment may be due to a new perception of the relationship between environment and development, to a perception that they are complementary rather than fundamentally opposed.

1. INTRODUCTION

In sociology, as in many other areas of western social and political thought, we are frequently confronted by the limitations in the way we and others conceptualise problems. For instance, if you are investigating the welfare state, if you are a collectivist, if you are not a wholehearted supporter of government intervention in welfare, then you must be against welfare, you must be an economic rationalist, a free market enthusiast. Viable options, for instance, the introduction of some mechanisms for greater consumer voice in the delivery of services, are thereby excluded from debate. Similarly, when discussing environmental protection, if you are for economic development, you must be opposed to environmental protection. The many ways in which economic development or economic rationality can complement environmental protection are thereby ignored (these might include market mechanisms, the use of ownership rights, price mechanisms, tax incentives, tradeable emission rights and enforcement incentives) (see Papadakis, 1993). One way of understanding these tendencies is examine the way in which we process information. Luhmann refers to the way in which symbols like 'money', 'power' and 'truth' are organised into 'binary codes' in order to simplify a complex environment.

These symbols are used, respectively, in the systems of the economy, politics and science. Along with the advantage of simplification come many disadvantages. Binary coding, by definition, makes it very difficult to conceive of alternative ways of perceiving situations and processing information. Luhmann perceives this problem in all areas of communication. He notes how binary codes entail, for instance, the use of the categories true/false in scientific discourse, of legal/illegal in the law and of government/opposition or progressive/conservative in politics (Luhmann, 1990:43). As I have suggested, social and political theorists are faced with a difficulty when trying to apply traditional ways of conceptualising social systems and traditional forms of logic (notably the use of adversarial logic and of rigid dichotomies and categories) to grasp new situations. Luhmann identifies some of these problems through his grasp of the operation of binary codes and 'self-referential systems'[2]. The binary codes used in self-referential systems impose huge constraints on dealing with new situations and on communication between existing subsystems. Hence, the economy as a self-referential system may only respond to the concern about environmental protection by 'translating' the language used to discuss concern about the environment into that of 'payments and prices': 'These codes are totalising in the sense that they exclude other possibilities of ordering information. A system can react to the environment *only* in terms of its code. For example, the binary code of the economy, payment/nonpayment, forces communications to be expressed in the language of prices and profits. This means that the economy can react to the environment, but as an autopoietically closed system it can do so only if it translates the language of nature into that of payments and prices.' (Fuchs, 1990:748).

The outcome of this 'structural blindness' of the economy to problems that cannot be translated into economic problems is illustrated by Luhmann: 'Even if, for example, fossil fuels deplete rapidly it may "still not yet" be profitable to switch to other forms of energy' (Luhmann, 1989:57 cited by Fuchs, 1990:748). The analysis of environmentalism and public debates surrounding this issue are rife with illustrations of the use of binary codes. Underlying this is the search for absolute truths and for 'certainty'. Many environmentalists have begun to recognise the problem of drawing sharp distinctions between economy and environment and have argued for (selective) economic growth - which may then lead to investment in more efficient and less environmentally damaging technology, or to a greater willingness by governments not to take short-cuts (like speeding up the process for granting permission to projects which carry a fairly high risk of damaging the environment). In Australia, as in many other countries, the environment has become a major political issue over the past decade. During that time several important changes have taken place. The green movement has gained political representation. The movement has become more organised and formalised. There have been subtle changes in the programs and concerns of green parties. Of particular relevance is the process of mutual adaptation between the (politically) most influential environmental groups and conventional political parties. Many of the concerns expressed by the greens have been accommodated by political elites, though there are still important areas of disagreement and conflict among these elites over the environment and development.

2. THE 1993 ELECTION CAMPAIGN

Environmentalists appeared to be confused by the 1993 Election Campaign. Phillip Toyne, former executive director of the Australian Conservation Foundation (ACF), noted that the community had become much more proenvironmentalist over the past five years. However, he was puzzled by the focus on the economy rather than the environment during the campaign[3]. I want to suggest that part of his confusion is due to thinking in the old manner, to positing a fundamental conflict between environment and economy. Some environmentalists understandably evoked this conflict during the election campaign, pointing out that the prime minister had continued to allow the export of woodchips from Australia and had ignored major issues and that his Environment Statement avoided contentious environmental problems like the establishment of firm targets for the reduction of greenhouse gases, the implementation of recommendations by the working groups on ecologically sustainable development and the introduction of legislation to protect endangered species. The Prime Minister simply used another form of green (or some would argue, brown) logic to counter this attack. He argued that the greens were not interested in 'brown' issues like the problems of salinity and soil degradation. These issues dominated his statement on the environment. Accompanying the statement was a program of measures which represented the latest electoral deal offered by the Australian Labor Party (ALP) for environmental protection. The package of measures was valued at $156 million over four years. Most of the funds (about $100 million) were to be spent on 'water quality initiatives', especially on cleaning up waterways like the Murray-Darling Basin. Other priorities included the development of conservation reserves that would be representative of plants, animals and habitats on this continent; the development of alternative and renewable sources of energy; the creation of a National Environment Information Database; assistance to companies that would serve as models for others in the area of waste prevention; the creation of a halon gas storage and recycling facility; and the management of tourism in environmentally sensitive areas. Another central feature of the campaign by the ALP was the argument that there is no fundamental conflict between development and the environment, that there is no need to choose between the two. Rather, the ALP government has suggested that 'we can, and must have both'. This theme was repeated in the Statement on the environment by the prime minister when he drew attention to the economic value of environmental protection[4]. He then praised the new cooperation between the Australian Council of Trade Unions (ACTU) and the

ACF over a 'green jobs' strategy. The speech stressed repeatedly the economic costs of environmental damage, particularly as a result of the condition of the Murray-Darling river system. All these statements represent the ability of major parties both in Australia and in other countries to steal the green agenda and to be highly adaptable and flexible about green concerns. Above all, it also represents their ability to deal with ambiguities and uncertainties in trying to formulate policies. It is also worth noting how the ALP has 'translated' concerns about environmental protection into the language of economics (see Luhmann, 1989 for an account of problems of communications between economic systems and environmental concerns). In political terms, the most interesting response to the prime minister's Statement came from the Liberal-National coalition which welcomed the proposed measures. The Liberals highlighted what they saw as a new emphasis by the ALP on cooperation rather than confrontation between the Commonwealth and the states (for instance, through the Intergovernmental Agreement on the Environment signed early in 1992). The Liberal spokesman on the environment, Jim Carlton, emphasised the 'convergence' in policies between the two parties. The Liberals were keenly aware of the success of the ALP over the past decade in promoting itself as the more responsible of the major parties on environmental issues. They therefore seized this opportunity to identify points in common with the ALP. The apparent similarities in policies were used by the Liberals as a basis for appealing to voters concerned about the environment and for trying to undo the links that had developed between environmentalists and the ALP. Among members of environmental groups, particularly among leading figures, there was also a reduction in emphasis on the ties with the ALP. However, this did not necessarily benefit the Liberal-National coalition. Rather, environmental groups expressed dissatisfaction with both major parties and refused to make any recommendation about where voters should direct their preferences. They also argued that the policies of the Liberal-National coalition posed a greater threat to the environment. They noted that the Liberals were in favour of extending projects for mining uranium, of reducing the fuel excise (which would lead to an increase in greenhouse gases), of handing back to the Northern Territory government the management of the Uluru and Kakadu national parks, of mining in Coronation Hill and of ending quotas on woodchips. Environmentalists were also concerned about proposals by both major parties for 'fast-tracking' development projects. In the case of the Liberals, a 12 month limit had been set on evaluating the environmental impact of a project. After that a project could proceed almost automatically if the evaluation was incomplete. To make its position clear, the ACF published an advertisement which condemned all the major parties. The ACF also noted some of the most significant differences between the environmental policies of the various political parties, illustrating these in graphic form with a system of 'ticks' and 'crosses' (see Figure 1).

Figure 1: Assessment of the major parties' environmental policies by the ACF

	Labour	Liberal/National	Democrats	Greens
Nuclear	●	x	√	√
Water quality	√	√	√	√
Air quality	●	x	√	√
National Parks	●	x	√	√
Endangered Species	●	●	√	√
Forests	●	x	√	√

√ means environmentally sound

● not good enough
x disaster
Source: Advertisement by the Australian Conservation Foundation, *Australian* 6-7 March 1993.

From this point on, groups like the ACF faced an uphill battle over shaping voters' perceptions. Rather than acknowledge the record of the ALP over the past decade, the ACF argued that it was simply not good enough. This negative picture, and the lack of recognition of the accomplishments by the ALP in government over the past decade, reflects the disappointment of environmentalists with the election campaign. Reflecting the disillusionment of environmental groups with the ALP, data from the 1990 and 1993 Australian Election Surveys (AES) show that the electorate as a whole was less likely in 1993 than in 1990 to identify with the ALP on environmental policy. (For more details about these studies see Gow et. al. 1990 and Jones et. al. 1993) The gap between the ALP and the Liberals appeared to have become narrower, with many more people feeling that there was no difference between the policies of the major parties. In 1990 most respondents (54%) nominated the ALP when asked which of the main parties came closest to their own view on environmental policy. In 1993 this figure had dropped dramatically - to only 34%. The percentage identifying with the Liberal-National coalition on environmental policy remained stable - 15% in 1990 and 17% in 1993. The big increase was in the number who felt that there was no difference between the major parties (from 17% to 32%). The decline in the number of people identifying with ALP policies on the environment applies not only to the population in general, but especially to people who are either members or supporters of environmental groups. Among those who were members of environmental groups, there was a sharp decline (from 82% to 59%) in those who identified with policies of the ALP (see Table 1). Among those who approved strongly of environmental groups there was also a decline in identification with policies of the ALP (from 73% to 49%). These changes correspond to the shift in emphasis during the 1993 election campaign, especially by the ALP.

Table 1 Identification with Party Policies on the Environment by Environmental Group Membership and Approval of Environmental Groups (percentages)

	Environmental Group Membership				Environmental Group Approval			
	Member		Not a member but have considered joining		Strongly approve		Approve	
Policy Preference[a]	1990	1993	1990	1993	1990	1993	1990	1993
ALP	82	59	74	51	73	49	57	33
Liberal-National	2	10	9	13	7	12	13	17
No difference	11	26	10	24	12	25	18	33
Don't know	6	6	7	11	9	14	12	17
Total	3	5	23	19	28	28	47	46
(N)	(55)	(126)	(413)	(521)	(512)	(762)	(857)	(1259)

[a] 'Whose policies - the Labor Party's or the Liberal-National Coalition's - would you say *come closest* to your own views on ... the environment'.

Sources: *Australian Election Studies*, 1990, 1993.

3. THE ENVIRONMENT AS A POLITICAL ISSUE

The enigma confronting prominent environmentalists about why the environment did not feature prominently during the election campaign can partly be resolved by analysing trends in opinion about the environment as a political issue. When asked how important the environment was in their decision about how to vote, in 1990 52% of the AES respondents indicated that it was extremely important (this represented a significant rise over the figure of 31% in 1987). However, in 1993 the figure had dropped 11 percentage points to 41 per cent. There was a comparable decline in the number of people who referred to the environment as the issue that had worried them and their families most over the previous 12 months: from 11 per cent in 1990 to only 4 per cent in 1993. One source of optimism for environmentalists lies in the follow-up question about which issues were likely to worry respondents and their families most 10 years from now: the environment, with 10.6 per cent, leapt into third place after unemployment (27.9 per cent) and health (13.7 per cent). As a political issue the environment was demoted in 1993. Yet, it was still on the agenda and likely to remain there for some time ahead. This claim is supported by data on the pattern of responses to several questions on environmental issues. In the 1993 AES respondents were asked about the urgency of various environmental concerns. The percentages indicating that these concerns were urgent resembled strongly the figures from the 1990 AES. In 1993 many people still regarded as fairly or very urgent, the problems of pollution (96 per cent), waste disposal (95 per cent), the Greenhouse effect (87 per cent), soil degradation (93 per cent), the destruction of wildlife (88 per cent), the logging of forests (76 per cent), uranium mining (58 per cent) and overpopulation (56 per cent). Pollution and the Greenhouse effect were still regarded as the most urgent problems (by 38 per cent and 16 per cent of respondents, respectively). These findings are reinforced by analysis of opinions about policies and about environmental groups. With respect to policy, in 1990, 79 per cent of respondents agreed that industry should be prevented from causing damage to the environment, even if this sometimes leads to higher prices. In 1993 the figure was almost identical (80 per cent). Similarly, in 1990 and 1993, 67 per cent of the sample agreed that the government should do more to protect the environment, even if this sometimes leads to higher taxes. There was also stability, at the aggregate level, in approval of environmental groups and movements (the percentage who either approved or strongly approved of them was 74 per cent in 1990 and 73 per cent in 1993) and in membership or potential membership (the percentage who indicated they were members of environmental groups was 3.0 per cent in 1990 and 4.5 per cent in 1993 and those who indicated that they were not members, but had considered joining was 22 per cent in 1990 and 18 per cent in 1993). All these findings support the proclamations by some prominent environmentalists that the environment remained an important issue in people's minds and they were still concerned enough about these issues to hold environmental groups in high regard and to join them. They were also prepared to continue supporting policies that implied tax rises and increases in prices in order to address environmental problems.

4. THE ENVIRONMENT VERSUS DEVELOPMENT?

It is worth noting again that continued support for environmental protection was recorded during a major economic recession. Yet, the most widely publicised explanation for the decline of the environment as a electoral issue was that people were too preoccupied about economic issues. Following his statement on the environment the prime minister was complemented for producing a statement that was 'right for the times': 'With the economy in general and unemployment in particular in the forefront of people's worries, the environment has moved into the electoral background. Concerns about the environment have not disappeared but to many people they rank behind anxiety about their, or their children's, employment prospects' (*Canberra Times*, 22 December 1992). This way of reasoning suggests that concerns about the environment are bound to clash with a preoccupation about the economy. This conflict has often been articulated by environmentalists who reject notions like sustainable development, arguing that living standards in wealthy countries are 'unsustainable', draw a sharp contrast between economic growth and the pursuit of human happiness and dramatise the consequences of further exploitation of the earth's resources. On the other hand, some developers regard environmentalism as an ideology which has replaced socialism and environmental groups as the vehicle for revolutionaries to subvert the political system. These groups apparently represent the interests of a new class of intellectuals, lawyers and bureaucrats who both profit from and wish to destroy the capitalist system (see Papadakis, 1993:chapter 3).

Before exploring how these issues affect electoral strategies, I should like to examine the argument about the relationship between a) perceptions of the economy and of environment and development on the one hand and b) opinions about environmental issues and policies as well as support for environmental groups on the other. Two variables, reflecting perceptions of the economy, are used as surrogates for attitudes towards development. Another two variables, views about uranium mining and about developing nuclear energy represent indicators of attitudes towards development and the environment. Beginning with the first pair of variables, both in 1990 and in 1993 participants in the AES surveys were asked a) how the financial situation in their household now compared with what it was 12 months ago and b) how they thought the general economic situation in Australia now compared with what it was a year ago. The other pair of variables was derived from two statements (about nuclear energy and uranium mining) which provided respondents with five options (ranging from strong agreement to strong disagreement). These two variables can be used as indicators of attitudes to development and the environment. Our analysis begins by examining the relationship between the two pairs of variables (representing views about the economy and views about the environment and development); we then explore the relationship between both pairs of variables and opinions about environmental policy and support for environmental groups.

Between 1990 and 1993 there was a small increase in the number of people who felt that nuclear energy is not a real necessity for the future (from 34 per cent to 40 per cent). By contrast, the number who were opposed to uranium mining in Australia dropped a little (from 34 per cent to 29 per cent). Perhaps surprisingly, the percentage of people who felt that the economic situation in the country had become worse over the previous 12 months declined from 74 per cent to 62 per cent between 1990 and 1993. Similarly, there was a reduction, from 54 per cent to 44 per cent, in the number of people who felt that the financial situation of their household had become worse over a 12 month period. In examining the relationships between these two sets of variables, we find that the relationship between perceptions of the financial situation in households and the development of nuclear energy was very weak ($r= .00$) in 1990 (Table 2). Yet, in 1993 it had become fairly significant ($r= -.05$). Among those who felt that their financial situation had worsened there was greater support in 1993 than in 1990 for nuclear energy. There was an equivalent change in the relationship between the financial situation of households and views about uranium mining. The change is even more pronounced with respect to views about the economic situation in the country as a whole. People who were concerned about this were much more likely in 1993 than in 1990 to regard nuclear energy as a real necessity for the future ($r= -.04$ and $r= -.13$, respectively). An equivalent change took place in views about uranium mining. However, if we focus instead on the relationship between views about the financial circumstances of households as well as the economic situation generally and views on tax increases in order to secure environmental protection there are virtually no changes in the strength of association. The relationship between opinions about price rises and views about the economic situation in general has also remained similar. The statistical significance of the relationship is weak both in 1990 and 1993. Similarly, there is a very weak relationship between views about the financial situation of households and opinions about price rises.

Table 2

Perceptions of the economy[a]/environment and development[b] and Opinions about environmental issues and policies[c]/Support for environmental groups/membership of environmental groups[d] (per cent)

	Nuclear Energy 1990	Nuclear Energy 1993	Uranium Mining 1990	Uranium Mining 1993	Low taxes versus spending 1990	Low taxes versus spending 1993	Low prices versus environmental protection 1990	Low prices versus environmental protection 1993	Approval of environmental groups 1990	Approval of environmental groups 1993	Membership of environmental groups 1990	Membership of environmental groups 1993
Financial situation of household	.00 (.90) -.12	-.05 (.01) -2.49	-.04 (.02) -1.81	-.06 (.00) -3.25	.09 (.00) 3.84	.08 (.00) 3.76	.03 (.29) 1.06	.01 (.50) .66	.08 (.00) 3.73	.010 (.00) 5.35	.09 (.00) 3.92	.06 (.00) 3.16
Economic situation in the country	-.04 (.08) -1.76	-.13 (.02) -6.71	-.07 (.00) -3.21	-.12 (.00) -6.32	-.10 (.00) 4.34	.11 (.00) 5.08	.01 (.66) .44	.01 (.56) .58	.12 (.00) 5.40	.11 (.00) 5.85	.09 (.00) 3.96	.10 (.00) 5.27
Nuclear Energy	.61 (.00) 33.84	.63 (.00) 42.48	.61 (.00) 33.84	.63 (.00) 42.48	-.17 (.00) -7.02	-.18 (.00) -8.65	-.13 (.00) -5.56	-.15 (.00) -7.50	-.22 (.00) -10.18	-.25 (.00) -13.76	-.20 (.00) -9.24	-.25 (.00) -13.71
Uranium Mining	.61 (.00) 33.84	.63 (.00) 42.48	—	—	-.24 (.00) -11.77	-.24 (.00) -9.99	-.15 (.00) -6.36	-.18 (.00) -8.88	-.34 (.00) -16.17	-.31 (.00) -17.46	-.30 (.00) -13.75	-.28 (.00) -15.58

Notes: The three sets of figures represent the Pearson Correlation, the significance level of the correlation (in brackets), and the T-Value.

a) 'How does the **financial situation of your household** now compare with what it was 12 months ago? And how do you think the **general economic situation in Australia** now compares with what it was a year ago? A lot better; a little better; about the same; a little worse; a lot worse.'

b) 'Here are some statements about general environmental concerns. Please say whether you strongly agree, agree, disagree or strongly disagree with each of these statements ... Nuclear energy is a real necessity for the future ... Australia should mine its uranium ... strongly agree, agree, neither agree nor disagree, disagree, strongly disagree,'

c) 'Which of these statements comes closest to your own views : Industry should be prevented from causing damage to the environment, even if this sometimes leads to higher prices OR Industry should keep prices down, even if this sometimes causes damage to the environment. Don't know, haven't thought much about it.'
'And which of these statements comes closest to your own views: Governments should do more to protect the environment, even if this sometimes leads to higher taxes for everyone OR Governments should keep taxes low, even if this sometimes means that they do less for government. Don't know, haven't thought much about it.'

d) See Table 1.

Sources: Australian Election Studies, 1990, 1993.

Turning to support for environmental groups, there is some evidence that among respondents who feel their financial situation has improved, approval of environmental groups has increased slightly between 1990 and 1993 (r= .08 and r=.10, respectively). There is also a slight increase in the strength of association between views about the economic situation in the country and membership of environmental groups (r = .09 in 1990 and r=.10 in 1993). Moving on to the second pair of variables, nuclear energy and uranium mining, which provide a more direct measure of views about development, we find that the two variables are strongly related to each other (r = .61 in 1990 and r= .63 in 1993). The relationship between these variables and specific measures of support for environmentalism (notably approval or membership of environmental groups) is generally very strong. In other words, support for environmentalism relates significantly to a reluctance to develop nuclear energy and to mine uranium. There is also a reasonably strong connection between these two measures of development and views about low taxes versus spending on environmental protection as well as low prices versus environmental protection. Those who favour low taxes or low prices are also more likely than others to favour the development of nuclear energy and uranium mining. However, there is again little change in the strength of these relationships between 1990 and 1993. Several other insights can be gained from the data. First, despite the severity of the recession, concern about environmental issues, support for environmental groups and for policy prescriptions involving financial sacrifices in order to protect the environment remain stable at the aggregate level. Second, those who are most concerned about their household situation as well as those who are concerned about the decline of the economy in general, are more likely, in 1993 than in 1990, to support the development of nuclear energy and uranium mining. Third, there have been virtually no shifts in the relationship between views about the economy in general as well as household financial circumstances and various policy proposals for tax increases and price rises to ensure environmental protection. This suggests that the overall link between economic circumstances and views about environmental protection is generally fairly weak. Fourth, there appears among certain groups to be a strengthening of ties between a positive view of the economic situation in the country and membership of environmental groups as well as between perceptions of household financial circumstances and approval of environmental groups. Fifth, there is a fairly strong connection between views about developing uranium mining or nuclear energy and support for environmentalism. Finally, this connection does not appear to have been influenced by the recent recession and represents a strong degree of continuity between the situation in 1990 and in 1993. The analysis therefore provides limited support for arguments positing a fundamental conflict between the objectives of economic growth and environmental protection. One of the most striking features of the 1993 election, which I shall elaborate more fully below, was the recognition, even by environmental groups, of the complementarity between economic and environmental goals. Despite this overall trend, there does remain a certain tension, at least at the ideological level, between the two objectives. This tension is articulated both by prominent environmentalists and by leaders in business and industry.

The next stage of the analysis examines this potential tension between development and the environment in two ways. First, it reports on the relationship between views about media coverage of environmentalists and perceptions of household financial circumstances as well as of the economic situation in the country as a whole. Second, it uses multivariate techniques to explore whether or not the bivariate relationships reported earlier (between, for instance, views about the economy and environmental issues) are still significant if we take other socio-economic factors into the account. In examining the relationship between views about media coverage of environmentalists and perceptions of household circumstances we find that those who regard their financial situation as a lot worse are somewhat more likely to agree that TV coverage is biased towards the greens ($r = -.08$) (Table 3). The relationship is stronger with perceptions of the economic situation in general. Respondents who perceive the economic situation in the country as a lot worse (49 per cent) are more likely than those who perceive it as a little better (32 per cent) to agree that television coverage of environmental issues is biased in favour of the greens ($r = -.16$). The strongest connections are between perceptions of TV bias and views about nuclear energy as a necessity or the desirability of mining uranium ($r = .27$ and $r = .32$, respectively). Turning to the multivariate analysis, we can now compare in a more rigorous manner the shift, between 1990 and 1993, in the relationship between a) environmentalism and b) perceptions of household financial circumstances, of the economy in general and of development and the environment (Table 4). The row of dependent variables reflects concerns about household finances, the economy and development. The independent variables represent two types. The first two variables indicate policy concerns and support for environmental issues: they are a) views on whether or not respondents are prepared to pay a higher price for goods in order to prevent industry from damaging the environment and b) approval of environmental groups. The remainder of the variables have been added as control variables. Most of these variables represent objective social location in terms of sex, age, level education, type of schooling and occupational status. The equation also includes a variable on self-rated class location.

Table 3 Perceptions of the Economy and Environment and Development, by Opinions about Television Coverage of Greens (percentages)

	Television bias[a]		
	Agree/ Strongly agree	Neither	Disagree/ Strongly disagree
Financial situation of Household			
A lot better	40	32	28
A little better	34	38	28
About the same	36	39	26
A little worse	41	33	26
A lot worse	46	32	22
Pearson Correlation: -.08	Sig: .000	T-value: -4.20	
Economic situation in the country			
A lot better	38	36	26
A little better	32	36	33
About the same	31	40	30
A little worse	35	39	27
A lot worse	49	32	19
Pearson Correlation: -.16	Sig: .000	T-value: -8.41	
Nuclear Energy a necessity			
Strongly agree	65	20	15
Agree	50	33	15
Neither	35	50	15
Disagree	33	32	35
Strongly disagree	26	30	44
Pearson Correlation: .27	Sig: 000	T-value: 14.93	
Uranium mining			
Strongly agree	66	20	14
Agree	51	33	16
Neither	32	47	21
Disagree	25	34	41
Strongly disagree	25	29	46
Pearson Correlation: .32	Sig: 000	T-value: 18.06	

Notes:

[a] 'Here are some statements about general environmental concerns. Please say whether you strongly agree, agree, disagree or strongly disagree with each of these statements ... TV coverage of environmental issues is biased in favour of the greens ... strongly agree, agree, neither agree nor disagree, disagree, strongly disagree.'

Sources: *Australian Election Study*, 1993.

Table 4: Opinions on the Environment and Other Factors as Predictors of Perceptions of the Economy, and Environment and Development[a]

	Household Finances 1990	Household Finances 1993	Situation in the Country 1990	Situation in the Country 1993	Nuclear Energy 1990	Nuclear Energy 1993	Uranium Mining 1990	Uranium Mining 1993
Price rises versus environmental protection	.00 (.88) .15	.00 (.87) .16	.06 (.01) 2.45	.02 (.57) .57	-.02 (.44) .78	-.03 (.20) -1.28	.00 (.88) .15	.00 (.93) -.09
Approval of environmental groups	.12 (.00) 4.51	.08 (.01) 2.87	.11 (.00) 4.12	.12 (.00) 4.32	-.17 (.00) -6.50	-.26 (.00) -9.79	-.24 (.00) -9.76	-.27 (.00) -10.52
Sex[b]	-.04 (.18) -1.34	-.06 (.04) -2.03	-.01 (.76) -.30	-.06 (.03) -2.19	-.07 (.01) -2.55	-.01 (.73) -.34	-.11 (.00) -4.54	-.08 (.00) -2.94
Age	.02 (.39) .87	-.09 (.00) -3.14	-.02 (.44) -.78	-.06 (.02) -2.27	.16 (.00) 6.15	.13 (.00) 5.03	.15 (.00) 6.00	.13 (.00) 4.99
Level of education	-.01 (.61) -.50	.05 (.11) 1.61	-.02 (.51) -.66	.03 (.31) 1.01	-.02 (.39) -.85	-.05 (.09) -1.70	-.01 (.60) -.53	-.04 (.08) -1.77
Schooling	-.04 (.12) -1.56	-.00 (.93) -.08	-.05 (.05) -1.96	-.03 (.27) -1.33	.08 (.00) 3.24	-.00 (.89) -.14	.03 (.18) 1.35	-.00 (.88) -.16
Occupational status	-.05 (.06) -1.89	-.01 (.63) -.48	-.10 (.00) -4.04	-.04 (.15) -1.43	-.03 (.20) -1.30	-.04 (.11) -1.60	-.04 (.11) -1.59	.02 (.39) .86
Self-rated class location	-.06 (.02) -2.34	-.01 (.54) -.61	.12 (.00) 4.63	.08 (.00) 2.96	.03 (.19) 1.30	-.08 (.01) -2.78	-.02 (.40) -.84	-.10 (.00) -3.74
R^2 F Sig.	.02 3.95 (.000)	.02 4.19 (.000)	.05 10.46 (.000)	.04 6.19 (.000)	.08 16.16 (.000)	.11 20.00 (.000)	.12 25.03 (.000)	.12 23.34 (.000)

[a] The cell entries are the standardised regression coefficient, the significance level (in parentheses) and the t-value.
[b] Coding of social location variables: Sex: male=1, female=0; Level of education: post-school training: none=1, yes=1; Schooling: government (state) school=1, Catholic=.5, other non-government=0; Occupational status: based on the Australia Standard Classification of Occupations (see Australia Bureau of Statistics 1986) and computed in the manner suggested by Jones (1989), values range from 0 to 100 with higher values representing higher status occupations; Self-perceived class location: middle=1, working=0.
Sources: *Australian Election Studies*, 1990 and 1993.

The first thing to note is the very close similarity between the equations for 1990 and for 1993 in the amount of variance explained. In three cases it is almost identical. Turning to the variable on low prices versus environmental protection, we find that this is a very poor predictor of views about household finances, of the economic situation in general and of the ideas about whether or not to develop nuclear energy and uranium mines. These findings provide a more robust test of the associations reported in Table 2, where we found a weak association between views on prices and views about household and general economic circumstances and a fairly strong association between views on prices and views on uranium mining and nuclear energy. The multivariate analysis confirms the findings in Table 2 of a significant association between approval of environmental groups and concerns about household finances and the economy in general. It also confirms the even stronger association between approval of these groups and views about developing nuclear energy and uranium mines. Though not reported here, a separate multivariate analysis using the same control variables was conducted to demonstrate that, as reported in Table 2, there has been a decisive shift in the relationship between perceptions of nuclear energy and of uranium mining and perceptions of the economic situation in general. This suggests that the recession has prompted some people to take more seriously the option of developing nuclear energy. Another factor which may account for the stronger relationship between views about the economic situation and opinions about nuclear energy is the Gulf War and the heightened awareness of the need for self-reliance based on this form of energy. However, there is no evidence of a strong shift in opinion in favour of nuclear energy. A more plausible explanation for the strong correlation between the two issues may be found in the publicity surrounding the campaign of the Liberal-National coalition and their endorsement of the nuclear option (see above, advertisement by the ACF). Among the other independent variables, sex was a much more significant predictor (in 1993 than in 1990) of views about the economic situation in general (with females more likely than before to express concern about this). However, sex declined as a predictor of views about nuclear energy (with females less likely than before to express the view that nuclear energy was a real necessity). These findings suggest that the recession may have influenced some females to be less anti-development than they had been in the past. Another striking finding, when examining the control variables, is the increase in the significance of age as a predictor of concerns about household finances and about the economy in general. This may reflect the strong impact of the recession on certain age groups, notably the younger ones. However, age as a predictor of views about nuclear energy and uranium mining remained very stable between 1990 and 1993, with young people more likely to oppose development than older ones. Occupational status and self-rated class were generally less significant in 1993 than in 1990 in predicting concerns about household finances and about the economy in general. However, self-rated class was much more significant in predicting views about nuclear energy and uranium mining, with the middle classes much more likely to oppose development in 1993 than in 1990.

The survey data indicate that there are some tensions, as far as people are concerned, between development and the environment. In particular, views about nuclear energy, illustrate how, during the recession, the connection between pro-development attitudes and economic circumstances has become stronger. However, there is no evidence that there has been a major shift away from concern about the environment, even during the recession. Though the survey data do not demonstrate this, messages from political elites and social movements during the election campaign stressed the complementarity of economic and environmental goals. The concluding section of this article presents evidence of the widespread support for notions of 'sustainable development' among social movements and among political elites.

5. SUSTAINABLE DEVELOPMENT

Although environmental groups like the ACF were disappointed with the major parties for neglecting the environment during the federal election, they also emphasised the importance of development. In the advertisement (reported earlier) in which the ACF took issue with the policies of the major parties half of the space was dedicated to the headline 'We can give you jobs'. Underneath this the first four sentences read: 'Jobs and the environment. It doesn't have to be one or the other. Because the environment doesn't cost jobs, it creates them. With the right priorities, we can have jobs and a healthy environment.' The emphasis on development and the environment represents a shift not only by environmentalists but by other groups in the community. During the election campaign both the National Farmers Federation and the Australian Council of Trade Unions welcomed the statement by the prime minister on the Environment. It is not surprising that farmers' representatives should welcome the specific measures for planting trees along the Murray river, for controlling feral animals and plants and for improving the water supply. However, it should also be noted that the government had been engaged in a constructive dialogue with both farmers and environmentalists for several years, particularly in the development of a National Soil Conservation Strategy. Similarly, there have been efforts to promote a dialogue between trade union organisations and environmentalists. Although there are still significant differences in emphasis between the two sides, the election campaign demonstrated that the ACTU was continuing to pay greater need to the ecological implications of economic development. Again, this reflects recent efforts by the ACF and the ACTU to promote the idea of 'green jobs'. The environment statement by the prime minister in December 1992 represents the attempt by policy makers to put this idea into practice. Though environmentalists were unhappy about the omission of certain issues from the statement, they could hardly question the measures themselves. Apart from gaining support of trade union and farmer organisations, the statement attracted the sympathy of business and industrial groups by underlining the importance of the burgeoning environment protection industry, of the market for treating and recycling waste, for reducing air pollution and for cleaning water. It is hardly surprising that the prime minister's Environment statement was also welcomed by the Liberal and National parties. In their joint policy statement, issued in February 1993, they affirmed their commitment to 'A Better Environment - *and* Jobs' (emphasis, as in the original document) (Liberal and National Parties, 1993). Their aim was to establish a new Department of Sustainable Development. The name is derived from the concept of sustainable development which was used by the United Nations World Commission on Environment and Development (1990). The report by this commission is cited in the policy statement by the Liberal and National parties to justify a focus on development and the environment. The two parties argued that: 'There is no reason why we cannot have both high economic growth and a sustainable environment'. They also acknowledged the 'valuable work and recommendations' of the ecologically sustainable development groups established by the Hawke Labor government - though they proposed a new plan to coordinate activities in this sphere. Analysis of policy platforms and of statements by prominent politicians and by social movement organisations reveals one of the most interesting features of the 1993 election campaign. It illustrates the consolidation and widespread acceptance of ideas about the complementarity of development and the environment. These ideas were actively promoted by the Labor government (for instance, through the ecologically sustainable development process and through various initiatives). The outcome was an unprecedented level of collaboration between interest groups, each of which laid different emphasis on aspects of policies for development and the environment. The recession and arguments about taxation played an important role in the demotion of the environment as a political issue in its own right during the 1993 election campaign. Paradoxically, the environment remains an issue of great concern to many people and is likely to remain firmly on the agenda. This chapter presents an explanation for this paradox: the reduction in emphasis on the environment as a very important electoral issue may largely be due to the perception shared by increasing numbers of people that development and the environment are not necessarily opposed and are often complementary to one another. The tendency towards adopting binary codes may have been open to some modification. This may represent a growing recognition both by environmentalists and policy makers that drawing sharp distinctions between economy and environment may not always be the desirable way through which to address the problems confronting modern societies.

REFERENCES

Australian Bureau of Statistics. (1986). *Australian Standard Classification of Occupations*. Canberra: Commonwealth of Australia.

Fuchs, S. (1990). 'Ecological Communication'. By Niklas Luhmann' (book review). *American Journal of Sociology* 96 (3):747-8.

Gow, D. et al. (1990). *Australian Election Study, Social Science Data Archives*. Canberra: The Australian National University.

Jones, R. et al. (1993). *Australian Election Study, Social Science Data Archives*. Canberra: The Australian National University.

Jones, F. (1989). 'Occupational prestige in Australia'. *Australian and New Zealand Journal of Sociology* 25:187-99.

Keating, P. (1992). *Speech by the Prime Minister, the Hon P J Keating, MP. Environment Statement Launch*. Adelaide, 21 December 1992, Commonwealth of Australia.

The Liberal and National Parties' Environment Policy. (1993). *A Better Environment and Jobs*. Liberal and National Parties of Australia, February 1993.

Luhmann, N. (1989). *Ecological Communication*. Cambridge: Polity Press.

Luhmann, N. (1990). *Political Theory and the Welfare State*, Berlin: De Gruyter.

Papadakis, E. (1994). 'Development and the Environment'. *Australian Journal of Political Science* 29 Special Issue, 66-80.

Papadakis, E. (1993). *Politics and the Environment: the Australian Experience*. Sydney: Allen and Unwin.

World Commission on Environment and Development. (1990). *Our Common Future*, Melbourne: Oxford University Press.

NOTES

1. I am grateful to Clive Bean for helpful comments on an earlier draft of this paper, to Gillian Evans and Peter Corrigan for assisting me in locating information about the 1993 election campaign and to the principal investigators of the Australian Election Studies (1990 and 1993) for data used in the analysis. I should also like to acknowledge institutional and financial support from the University of New England, the Australian Research Council (Project on Environmentalism, Public Opinion and the Media) and from the Australian National University, Research School of Social Sciences (Project on Reshaping Australian Institutions). This chapter is based, with modifications, on a paper published in a Special Issue of the *Australian Journal of Political Science* (see Papadakis, 1994).

2. For instance, the term self-referential can be useful in understanding how the political system produces and reproduces itself:
'Whatever can become politically relevant results from a connection with whatever already possesses political relevance. Whatever counts politically reproduces itself. And this occurs by encompassing and absorbing interests from the social environment of the political system. Politics conditions its own possibilities - and apparently becomes sensible thereby to what its environment offers or requires. It is not understood adequately as a closed or an open system. It is both at the same time.The difficulties that theory-formation and on-going scientific research encounter here are rooted in their object. We will subsume them under the concept of "self-referential system."A system is called self-referential that *produces and reproduces* the elements - in this case political decisions - out of which it is *composed itself*' (Luhmann, 1990:39-40).

3. '... what is truly remarkable is the total absence of debate on the environment so far in the campaign, marking an enormous contrast with the last three federal elections. This is surprising because there is no evidence to suggest that it has ceased to be an important issue' (*Canberra Times*, 1 March 1993).

4. 'For too long the myth that jobs and environmental protection are incompatible has tended to govern our thinking. In truth, while some conflicts will inevitably remain for Governments and communities to resolve, it is increasingly evident that the economic sustainability of Australia is dependent on the environmental sustainability of Australia. ... the drive for environmentally friendly industries and the protection of our environment is part of the economic drive - part of the international competitive drive in which Australia is engaged.' (Speech by the prime minister, Paul Keating, Adelaide, 21 December 1992)

8. Global Environmental Concern: a Challenge to the Post-materialism Thesis

Riley E. Dunlap
Angela G. Mertig

Abstract[1]

It is widely assumed that public concern for environmental quality is dependent on affluence, and is therefore stronger in wealthy nations than in poor nations. This assumption is tested in this chapter via results from a 1992 international survey conducted by the George H. Gallup International Institute that obtained data on a wide range of environmental perceptions and opinions from citizens in 24 economically and geographically diverse nations. Aggregate, national-level scores for a variety of measures of public concern for environmental quality were created and correlated with per capita gross national product. Although the results vary considerably depending upon the measure, overall national affluence is more often negatively rather than positively related to citizen concern for environmental quality--contradicting conventional wisdom.

1. INTRODUCTION

Conventional wisdom has long held that concern about environmental quality is limited primarily to residents of the wealthy, highly industrialized nations located primarily in the Northern hemisphere, as residents of the poorer, non-industrialized nations are assumed to be too preoccupied with economic and physical survival to be concerned about environmental problems (e.g., Beckerman, 1974:89). This assumption was apparent in media reports concerning the 1992 UN Conference on Environment and Development in Rio de Janeiro (e.g., Elmer-Dewitt, 1992) and, interestingly, it has received support via social-scientific analyses of environmentalism. When accounting for the development of green parties and public support for environmental protection, political scientists typically argue that environmentalism stems from the emergence of 'post-materialist values' (Inglehart, 1990). Such values, it is argued, have resulted from post-World War II affluence in the industrialized nations, and represent a growing concern for quality of life over economic welfare among younger generations that take the latter for granted. Underlying the materialist/post-materialist distinction is the assumption of a hierarchy of human needs and values as suggested by Maslow; thus, it is not surprising that this perspective is compatible with that of psychologists who argue that environmental concern is unlikely to develop in societies where basic human needs are poorly met (Leff,

1978:50)[2]. Similarly, sociologists typically view environmentalism as an exemplar of the 'new social movements' (e.g., the environmental, peace, antinuclear and feminist movements) that have arisen within the wealthy, industrialized societies to pursue lifestyle and quality-of-life goals rather than economic interests (Buttel, 1992). Finally, economists widely regard environmental quality as a 'luxury good' that is likely to be of concern only to those who do not have to worry about food, housing and economic survival (Baumol and Oates, 1979). Since the emergence of post-materialist values (and higher-order needs) and the development of the new social movements that espouse them are viewed as dependent upon widespread, sustained affluence, these convergent theoretical perspectives all suggest that residents of the poor, non-industrialized nations will be less concerned about environmental problems and less supportive of environmental protection than are their counterparts in the wealthy nations. The early emergence of environmentalism and green parties in the industrialized world (primarily North America and Western Europe) lends support to the above perspectives, as did the wary reaction to the 1972 UN Conference on the Human Environment in Stockholm among non-industrialized nations. However, the more enthusiastic participation of the developing nations in the 1992 'Earth Summit' in Rio, a followup to the previous UN conference (Haas, et al., 1992), and especially the gradual emergence of environmental activism throughout much of the non-industrialized world (Durning, 1989; Finger, 1992), clearly pose challenges to conventional wisdom. It could be argued, however, that these phenomena are atypical, and represent only the responses of government elites or small numbers of activists in the poorer nations. In other words, conventional wisdom concerning rich-poor national differences in environmental concern might hold true at the level of the general public. Because existing surveys of public opinion toward environmental issues tend to be confined to North America and Western Europe (which vary only moderately in levels of affluence), very little is known about the general public's views of environmental issues in non-industrialized nations. Consequently, it has heretofore not been possible to test the assumption that residents of poor nations are less environmentally concerned than are their counterparts in the wealthy nations[3]. This situation has changed as a result of a recent international environmental opinion survey sponsored by the George H. Gallup International Institute.

2. METHODOLOGY

Sampling and Data Collection

The Health of the Planet (HOP) survey, coordinated by the George H. Gallup International Institute, was conducted in 24 economically and geographically diverse nations by members of the worldwide network of Gallup affiliates. The selection of countries was dependent upon the existence of a Gallup affiliate (or willing partner) and the availability of adequate funding. While the poorer, less-economically developed nations (especially African nations) are consequently underrepresented, the intent was *not* to conduct a worldwide survey whose results could be generalized to the entire world (an unrealistic goal). Rather, the goal was to compare the public's views of environmental issues across a wide range of nations, in terms of both geographic location and level of economic development. As shown in Table 1, the 24 nations covered in the HOP include six classified as 'low' income nations, seven as 'medium' income nations, and eleven as 'high' income nations by the World Bank on the basis of per capita gross national product (World Bank, 1992). Comparing citizens' views of environmental issues across such a wide range of nations should provide a reasonable test of the assumption that public concern for environmental quality is much stronger among the wealthy, industrialized nations than among their less-economically developed counterparts.

Table 1. Per Capita GNP and Sample Size by Nation[a]

Country/Economic Level	Per capita GNP	Sample Size
Low income		
Nigeria	290	1,195
India	350	4,984
Philippines	730	1,000
Turkey	1630	1,000
Poland	1690	989
Chile	1940	1,000
Middle income		
Mexico	2490	1,502
Uruguay	2560	800
Brazil	2680	1,414
Hungary	2780	1,000
Russia	3200	964
Portugal	4900	1,000
Korea (Rep.)	5400	1,500
High income		
Ireland	9550	928
Great Britain	16100	1,105
Netherlands	17320	1,011
Canada	20470	1,011
United States	21790	1,032
Denmark	22080	1,019
Germany (West)	22320	1,048
Norway	23120	991
Japan	25430	1,434
Finland	26040	770
Switzerland	32680	1,011

[a] Per Capita GNP based on World Bank (1992) categorization.

Each affiliate was responsible for translating the questionnaire into the appropriate language(s) for their nation, and then the Gallup International Institute had them 'back-translated' into English to insure comparability. The surveys were conducted via face-to-face, in-home interviews (thus minimizing problems of illiteracy), and all were completed during the first quarter of 1992. Nationally representative samples were used in all nations but India, where rural areas and regions experiencing terrorism were underrepresented (and thus caution must be used in generalizing the results to the nation as a whole). As can be seen in Table 1, sample sizes ranged from a low of 770 in Finland to nearly 5000 in India, and most were within the 1000 to 1500 range--yielding results that should have margins of error of approximately 3 percent for the respective national populations.

Measurement of Variables

The HOP survey examined a wide range of topics related to environmental issues (Dunlap, et al., 1993a; 1993b), and drew upon existing research to insure that key substantive domains and theoretical conceptualizations of environmental attitudes were employed (Dunlap, 1995; Ester and Van der Meer, 1982; Van Liere and Dunlap, 1981). In particular, we went beyond existing sociological and political science studies that have focused narrowly on 'environmental activism' (e.g., Inglehart, 1990) by employing a wide range of indicators of public concern about environmental quality. We will examine several dealing with the perceived seriousness of environmental problems (at various geographical levels), personal concern about these problems, and support for environmental protection to provide a detailed examination of conventional wisdom about differences between residents of poor and wealthy nations. Some of our variables are single items, while others are multi-item composites. The variables, item(s) on which they are based, and coding procedures are reported in detail in Table 2 (frequency distributions for all items are reported in Dunlap, et al., 1993a).

Table 2. Description of Variables

Perceived Seriousness of Environmental Issues in Nation
Question wording: 'I'm going to read a list of issues and problems currently facing many countries. For each one, please tell me how serious a problem you consider it to be **in our nation**... Environmental Issues.' *Coded as*: 1=Not at all serious, 2=Not very serious, 3=Somewhat serious, 4=Very serious. (Not sure, don't know and refused were deleted from analysis.)

Perceived Seriousness of Environment Relative to Other National Problems
Question wording: Same as prior item. Other National Problems: Hunger and homelessness: Crime and violence; Poor health care; The high cost of living; Racial, ethnic, or religious, prejudice and discrimination. *Coded as*: Score for prior item minus the mean rating of the other five national problems.

Personal Concern about Environmental Problems
Question wording: 'How concerned are you personally about environmental problems?' *Coded as*: 1=Not at all, 2=Not very much, 3=A fair amount, 4=A great deal. (Not sure and refused were deleted from analysis.)

Perceived Quality of Nation's Environment
  Question wording: 'Overall, how would you rate the quality of the environment in our **nation**?' *Coded as*: 1=Very good, 2=Fairly good, 3=Fairly bad, 4=Very bad. (Not sure, don't know and refused were deleted from analysis.)

Perceived Quality of Community Environment
  Question wording: 'Overall, how would you rate the quality of the environment here in your local **community**?' *Coded* same as prior item.

Perceived Quality of World's Environment
  Question wording: 'Overall, how would you rate the quality of the environment of the **world** as a whole?' *Coded* same as prior item.

Perceived Health Effects of Environmental Problems at Present
  Question wording: 'Please tell me - how much, if at all, you believe environmental problems now affect your health?' *Coded as*: 1=Not at all, 2=Not very much, 3=A fair amount, 4=A great deal. (Not sure, don't know and refused were deleted from analysis.)

Perceived Health Effects of Environmental Problems in the Past
  Question wording: 'Please tell me - how much, if at all, you believe environmental problems affected your health in the past - say 10 years ago?' *Coded* same as prior item.

Perceived Health Effects of Environmental Problems in the Future
  Question wording: 'Please tell me - how much, if at all, you believe environmental problems will affect the health of our children and grandchildren - say over the next 25 years?' *Coded* same as prior item.

Average Perceived Seriousness of Six Community Environmental Problems
  Question wording: 'Here is a list of environmental problems facing many **communities**. Please tell me how serious you consider each one to be here **in your community**... Poor water quality; Poor air quality; Contaminated soil; Inadequate sewage, sanitation and garbage disposal; Too many people, overcrowding; Too much noise.' *Coded as*: 1=Not at all serious, 2=Not very serious, 3=Somewhat serious, 4=Very serious. (Not sure and refused were deleted from analysis and individuals with fewer than four items with valid data were deleted from analysis.) *Standardized alphas* for each country range from .74 (Uruguay) to .89 (Finland, Norway).

Average Perceived Seriousness of Seven World Environmental Problems
  Question wording: 'Now let's talk about the **world as a whole**. Here is a list of environmental issues that may be affecting the world as a whole. As I read each one, please tell me how serious a problem you **personally** believe it to be in the **world**... or you don't know enough about it to judge?... Air pollution and smog; Pollution of rivers, lakes, and oceans; Soil erosion, polluted land, and loss of farmland; Loss of animal and plant species; Loss of the rainforests and jungles; Global warming or the "greenhouse" effect; Loss of ozone in the earth's atmosphere.' *Coded* same as prior item. (Can't judge and refused were deleted from analysis and individuals with fewer than five items with valid data were deleted from analysis.) *Standardized alphas* for each country range from .72 (India) to .90 (Mexico).

Average Level of Support for Six Environmental Protection Measures
  Question wording: 'Here are some actions our government could take to help solve our nation's environmental problems. **Keeping in mind that their costs associated with these actions**, please tell me - for each one I read - whether you would strongly favor, somewhat favor, somewhat oppose, or strongly oppose this action... Make stronger environmental protection laws for business and industry; Make laws requiring that all citizens conserve resources and reduce pollution; Provide family planning information and free birth control to all citizens who want it to help reduce birth rates; Support scientific research to help find new ways to control pollution; Limit exports of our natural resources to other nations; Ban the sale of products that are unsafe for the

environment.' *Coded as*: 1=Strongly oppose, 2=Somewhat oppose, 3=Somewhat favor, 4=Strongly favor. (Not sure and refused were deleted from analysis and individuals with fewer than four items with valid data were deleted from analysis.) *Standardized alphas* for each country range from .46 (Nigeria) to .86 (Poland), with the majority ranging from .60 to .69.

Preferred Trade-Off Between Environmental Protection and Economic Growth
Question wording: 'With which one of these statements about the environment and the economy do you **most** agree? Protecting the environment should be given priority, even at the risk of slowing down economic growth; Economic growth should be given priority, even if the environment suffers to some extent.' *Coded as*: 1=Economic growth given priority, 2=Equal priority (if volunteered), 3=Environment given priority. (Not sure was coded as equal priority and refused was deleted from the analysis.)

Willingness to Pay Higher Prices for Environmental Protection
Question wording: 'Increased efforts by business and industry to improve environmental quality might lead to higher prices for the things you buy. Would you be willing to pay higher prices so that industry could better protect the environment, or not?' *Coded as*: 1=Not willing, 2=Not sure/Don't know, 3=Yes, willing. (Refused was deleted from the analysis.)

For each of the 14 environmental variables, we created national-level aggregate scores for each nation by computing the national mean of all responses. (Means and standard deviations for all variables are reported for individual nations in the Appendix.) We did this because our concern is with *national-level* differences in environmental concern--specifically, in determining whether levels of concern are positively correlated with national affluence as commonly assumed. To measure the latter we used per capita gross national product, as reported in Table 1. In future analyses we plan to examine a variety of indicators of national affluence and quality of life, but since per capita GNP is the most widely used measure of national affluence it provides the most direct test of conventional wisdom[4]. Eventually we will also conduct individual-level analyses that examine the relationships between socio-economic status and environmental concern within individual nations, as well as contextual analyses to explore possible variation in the social bases of environmental concern across nations, but such analyses are beyond the scope of this chapter.

Data Analysis

Although conventional wisdom and social science theories hold that public concern for the environment is higher in wealthy nations than in poor nations, the precise nature of the relationship expected between national affluence and citizens' levels of environmental concern has not been specified. While the implicit assumption seems to be that such concern increases more or less linearly with affluence, some have argued for a more dramatic increase once a nation becomes highly industrialized--i.e., limited variation among poorer nations and among wealthy nations, but a vast difference in citizen concern between the two types of nations (Brechin and Kempton, 1994). Given the ambiguous nature of the expected positive relationship between national affluence and citizens' environmental concern, we conducted two types of analyses. First, we computed

Pearson's correlation coefficients between per capita GNP and each of the 14 dependent variables for the 24 nations, since at a minimum the conventional view leads us to expect positive relationships between affluence and environmental concern[5]. Second, we also ran scattergrams for each correlation, in order to visually examine whether the observed relationships were obviously non-linear. We report both correlation coefficients and scattergrams in this chapter. Some analysts have argued for using the logarithm of per capita GNP, on the assumption that differences of a few hundred dollars, for example, mean more among poorer nations than among wealthy nations (Brechin and Kempton, 1994). We therefore conducted parallel analyses using both GNP/capita and log of GNP/capita. We report correlation coefficients for both of these affluence variables, but since the results are very similar we will only show the scattergrams for GNP/capita in the interest of space. And finally, because we know that the sample for India is unrepresentative of rural areas, and because we have to employ a rough estimate of per capita GNP for the Republic of Russia, we also repeated the analyses without India and Russia. However, since the deletion of these two countries had no substantial effect on any of the reported correlations, and certainly not on our overall conclusions, we will not report the results of this analysis.

3. RESULTS

Perceived Seriousness and Personal Concern

A criticism of U.S. environmental polls is that they often fail to examine environmental issues within the context of other important issues, and thus elicit misleadingly strong indications of public concern about environmental quality (Dunlap and Scarce, 1991). To deal with the problem we began the HOP survey by asking a couple of questions about national level problems *before* respondents were aware that the major focus of the survey was on environmental issues. We first asked an open-ended item, 'What do you think is the most important problem facing our nation today?' The results are not used in the paper because this item yields a dichotomous variable (volunteered environmental problems or not) inappropriate for use in correlational analysis, but are reported elsewhere (Dunlap, et al., 1993a; 1993b). Overall, residents of the wealthier nations are somewhat more likely to volunteer environment as their nation's 'most important problem,' but the results reveal that environmental quality is a surprisingly salient public issue throughout the 24 nations--and thus a meaningful topic for opinion surveys in all of them.

Table 3. Correlations between National-Level Measures of Environmental Awareness and Concern and Per Capita GNP and Log of Per Capita GNP

Variable	GNP/Capita	Log GNP/Capita
Perceived Seriousness of Environmental Issues in Nation	-0.18	-0.13
Perceived Seriousness of Environment Relative to Other National Problems	0.69***	0.73***
Personal Concern about Environmental Problems[a]	-0.49*	-0.48*
Perceived Quality of Nation's Environment	-0.56**	-0.47*
Perceived Quality of Community Environment	-0.61**	-0.55**
Perceived Quality of World's Environment	0.41*	0.63**
Perceived Health Effects of Environmental Problems at Present	-0.69***	-0.66***
Perceived Health Effects of Environmental Problems in the Past	-0.26	-0.33
Perceived Health Effects of Environmental Problems in the Future	-0.54**	-0.47*
Average Perceived Seriousness of Six Community Environmental Problems	-0.52**	-0.58**
Average Perceived Seriousness of Seven World Environmental Problems	0.05	0.30
Average Level of Support for Six Environmental Protection Measures	-0.79***	-0.67***
Preferred trade-Off Between Environmental Protection and Economic Growth	0.55**	0.73***
Willingness to Pay Higher Prices for Environmental Protection	0.50*	0.66***

[a] Poland deleted *p<.05 **p<.01 ***p<.001

We then gave respondents a list of six national-level problems and issues, including environment, and asked them to rate the seriousness of each one on a four-point scale. As shown in Table 3, there is a slightly negative (albeit insignificant) relationship (r = -.18, n.s.) between GNP/capita and **Perceived Seriousness of Environmental Issues in Nation**--the opposite of what conventional wisdom predicts. It is also important to emphasize that citizens in all 24 nations tend to see environmental problems as at least 'somewhat serious,' as is apparent in Figure 1.

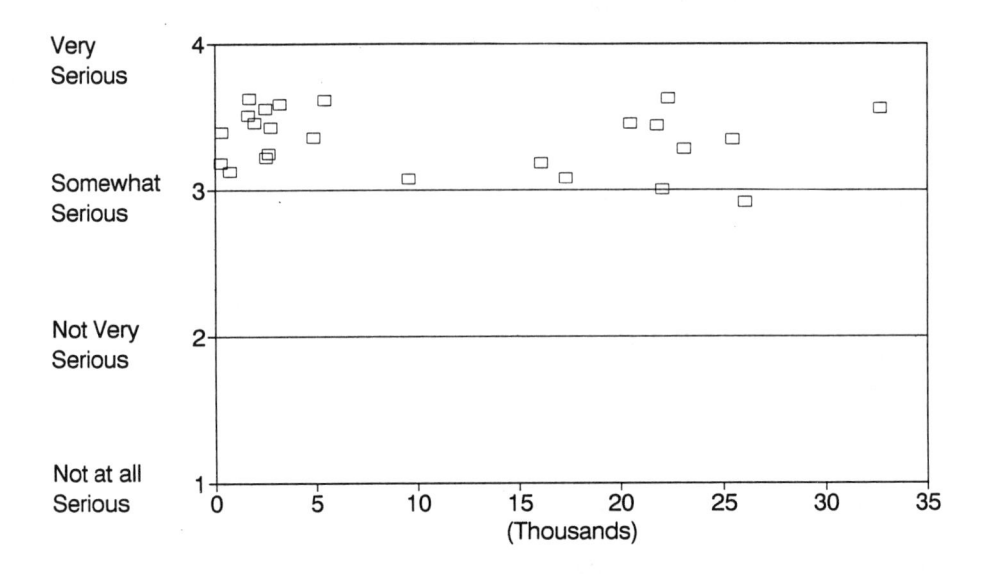

Figure 1. Perceived Seriousness of Environmental Issues in Nation By Per Capita GNP (1990 US Dollars)

Since all six national problems/issues were rated as more serious in the poorer nations, we created a new variable to measure the perceived seriousness of environment *relative* to the other problems by subtracting the mean rating for these five from that given to environment. The correlation between GNP/capita and **Perceived Seriousness of Environment Relative to Other National Problems** is strongly positive and significant ($r = .69$, $p < .001$), indicating that environmental issues are rated as *relatively* less serious in low-income nations. It should be noted, however, that even in a majority of the poor countries (i.e., those with GNP/capita of $5,000 or less) environment is rated *above average* in seriousness as illustrated in Figure 2.

Which of the above two variables is the better indicator of the perceived seriousness of environmental issues is debatable, but the results seem clear: Residents of low-income nations are slightly more likely to rate environment as a serious problem, but significantly less likely to rate it as serious relative to other national problems, than are their counterparts in high-income nations. Citizens in all 24 nations, however, tend to see the environment as a relatively serious problem.

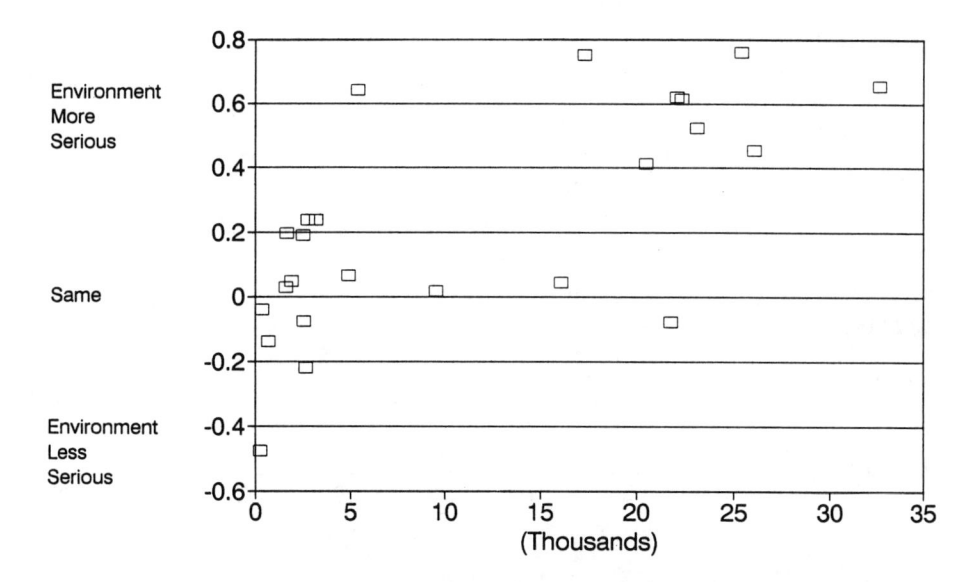

Figure 2. Perceived Seriousness of Environment Relative to Other National Problems By Per Capita GNP (1990 US Dollars)

A more direct indicator of personal concern about environmental quality was used in the HOP after we had informed respondents of our particular interest in environmental issues (as explained below). We employed a standard measure of environmental concern by asking respondents directly, 'How concerned are you personally about environmental problems?' and allowing them to respond on a four-point scale. (In Poland the item was incorrectly translated as 'How much attention do you give to environmental problems?' and we have deleted the Polish data for this variable.) The correlation between **Personal Concern about Environmental Problems** and per capita GNP is surprisingly the opposite of that predicted by conventional wisdom ($r = -.49$, $p < .05$), as residents of low-income nations tend to express higher levels of concern about environmental problems than do those of high-income nations. This negative relationship is quite obvious in Figure 3, which shows that the mean levels of national concern (while they are relatively high in most countries) declines with levels of national affluence.

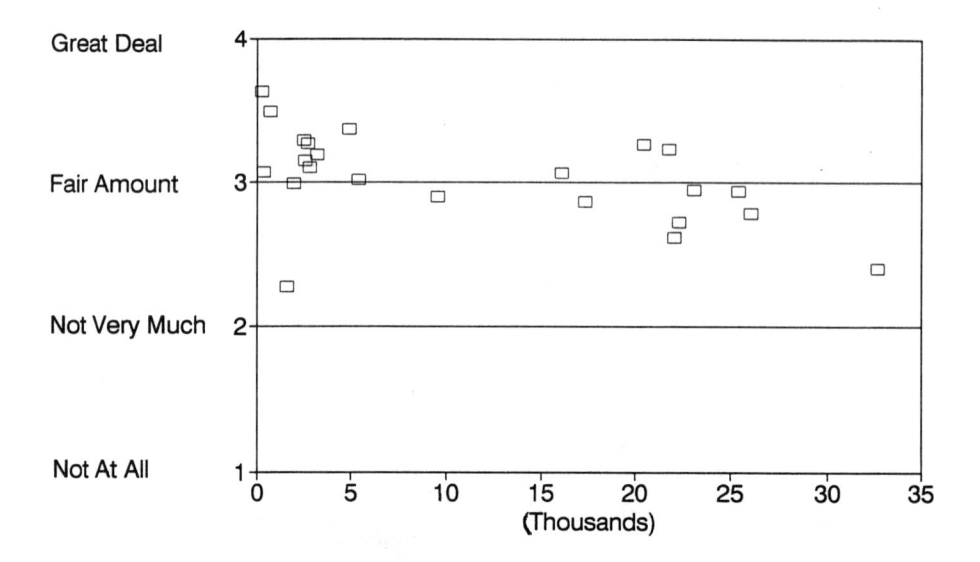

Figure 3. Personal Concern about Environmental
    Problems By Per Capita GNP (1990 US Dollars)

Ratings of Environmental Quality

To get a sense of why people around the world are concerned about environmental problems, the HOP survey asked them to rate the quality of the environment at the national, local and world levels. This question also provided a transition to the specifically environmental focus of the questionnaire (it was asked immediately *prior to* the 'personal concern' item just discussed) and as a means of clarifying exactly what we meant by 'environment.' After defining what we meant by environment (see Dunlap, et al., 1993a; 1993b) we asked respondents to 'rate the quality of the environment' for their nation, then for their local community and finally for the world as a whole. Consistent with prior research (e.g., Murch, 1971), we found that in most nations the world environment was rated the worst and the community environment the best, with the national environment in between (as can be discerned from the relevant figures discussed below).

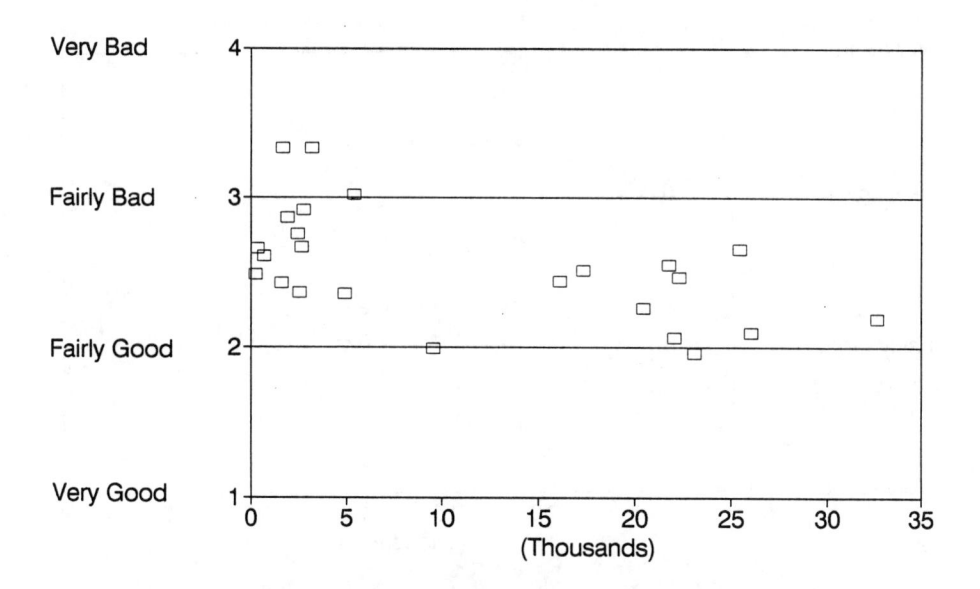

Figure 4. Perceived Quality of Nation's Environment
By Per Capita GNP (1990 US Dollars)

The pattern of correlations between per capita GNP and ratings of environmental quality at these three geographical levels is quite interesting. The relationships are significantly negative for both the national level ($r = -.56$, $p < .01$ for **Perceived Quality of Nation's Environment**) and community level ($r = -.61$, $p < .01$ for **Perceived Quality of Community Environment**), indicating that the poorer the nation the more likely residents are to rate their local and national environments badly. The relationship for the national environment is shown in Figure 4, and that for the community level in Figure 5. In contrast, the relationship is reversed for **Perceived Quality of World's Environment**, as residents of high-income nations are more likely to see the world environment as being in bad shape than are those in low-income nations ($r = .41$, $p < .05$). This relationship is shown in Figure 6.

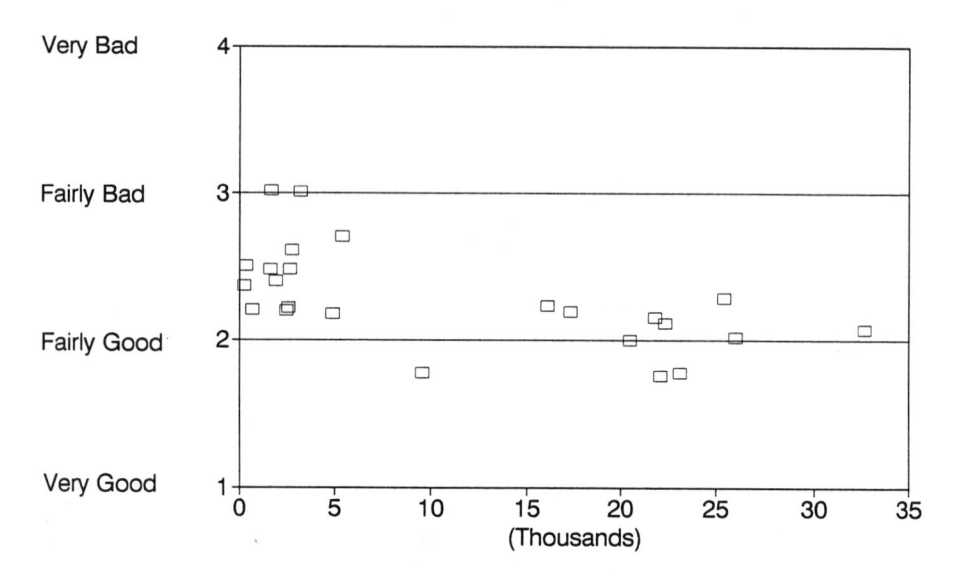

Figure 5. Perceived Quality of Community Environment
By Per Capita GNP (1990 US Dollars)

The results provide another indication that conventional wisdom about national differences in environmental concern is misguided. Citizens of poorer nations not only rate their community environment as significantly worse than do their counterparts in richer nations (which may not be so surprising, given the high levels of pollution found in many Third World cities) but the same pattern holds for the national environment as well. It is only when we focus on the worldwide environment that we find residents of wealthy nations more likely to offer negative ratings. At a minimum, these results reflect the importance of specifying the geographical referent when discussing differences in public concern about environmental problems across nations.

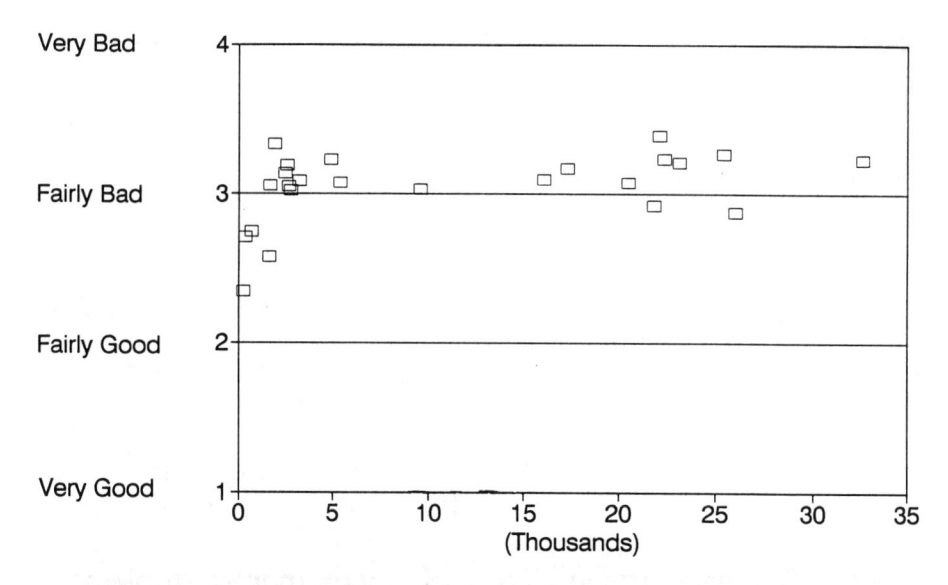

Figure 6. Perceived Quality of World's Environment
By Per Capita GNP (1990 US Dollars)

Environmental Problems as Health Threats

At the time of the 1972 Stockholm conference, environmental problems were widely viewed as primarily aesthetic issues or threats to the beauty of nature. This likely contributed to the cool reception given to environmental protection by the poorer nations at that time (see, e.g., Founex Report, 1972). A major change in the past two decades has been a growing awareness that environmental problems pose threats to human health, and this appears to have contributed to the rising concern over environmental deterioration observed in the U.S. and Europe (see, e.g., Dunlap and Scarce, 1991).

To investigate the perceived health effects of environmental problems in the HOP, we asked respondents to indicate, 'How much, if at all, you believe environmental problems *now* affect your health?' and then asked them if it affected their health *10 years ago* and if they thought it would 'affect the health of your children and grandchildren--say over the *next 25 years.*' We found a strong increase in reported health effects over time. While majorities in only four nations report that environmental problems affected their health at least 'a fair amount' ten years ago, majorities in 16 nations report such health affects at present and majorities in all 24 countries expect them over the next 25 years.

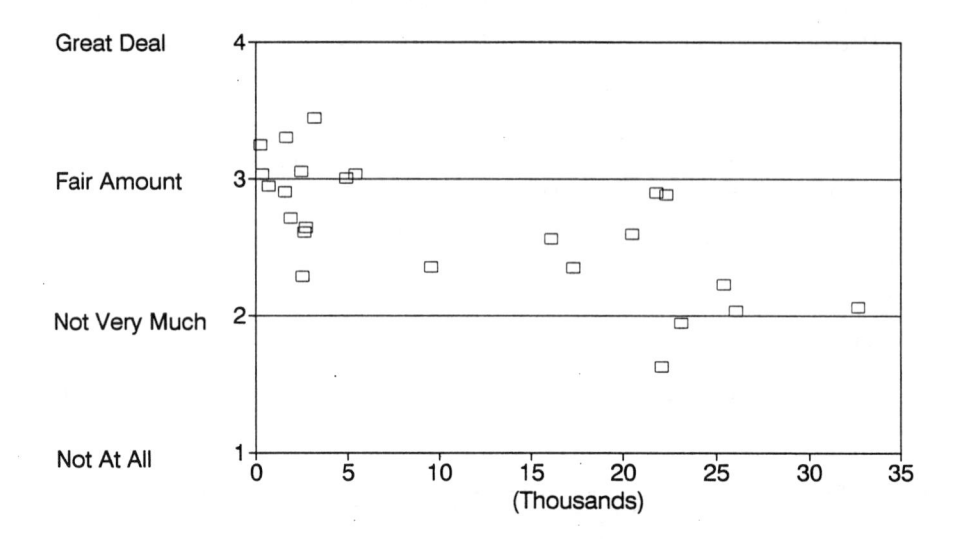

Figure 7. Perceived Health Effects of Environmental Problems at Present By Per Capita GNP (1990 US Dollars)

In all three cases--**Perceived Health Effects of Environmental Problems at Present**, in **the Past**, and in **the Future**--the relationship between perceiving health effects from environmental problems and national affluence is negative, indicating that residents of poorer nations are more likely than their counterparts in the wealthier nations to perceive environmental problems as threats to their health (see Figures 7, 8 and 9). Although the relationship is not statistically significant for past health effects (r = -.26, n.s.), it is significant for both present health effects (r = -.69, p<.001) and for future health effects (r = -.54, p<.01).

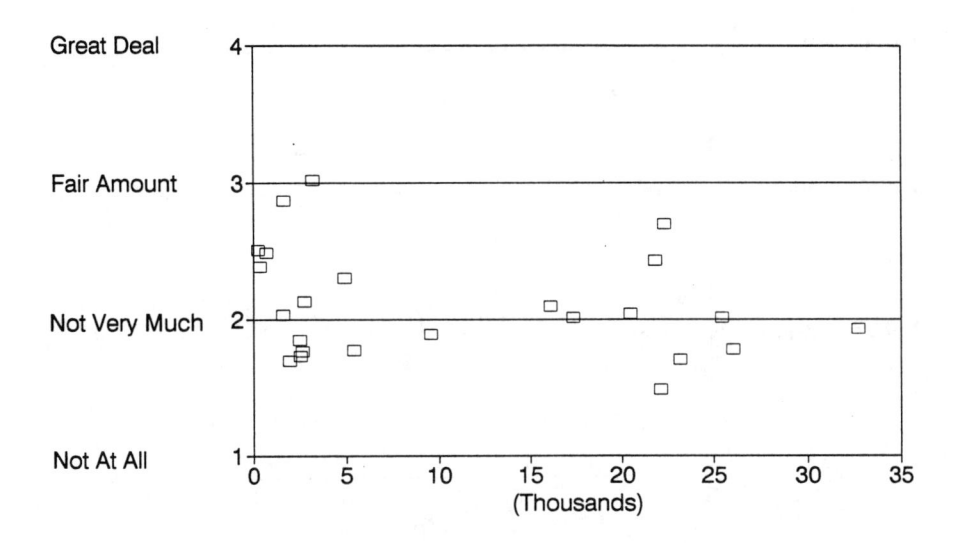

Figure 8. Perceived Health Effects of Environmental Problems in the Past By Per Capita GNP (1990 US Dollars)

The fact that residents of poorer nations are significantly more likely to feel that their present (and future) health is being (or will be) negatively affected by environmental problems provides insight into their previously noted higher levels of personal concern about such problems. Conventional wisdom should acknowledge that environmental problems are no longer viewed as just a threat to quality of life, but as a fundamental threat to human welfare. This threat is especially great in poor nations, where people often depend directly on the immediate environment for sustenance (food, water, fuel and building materials) and environmental degradation therefore threatens their very survival.

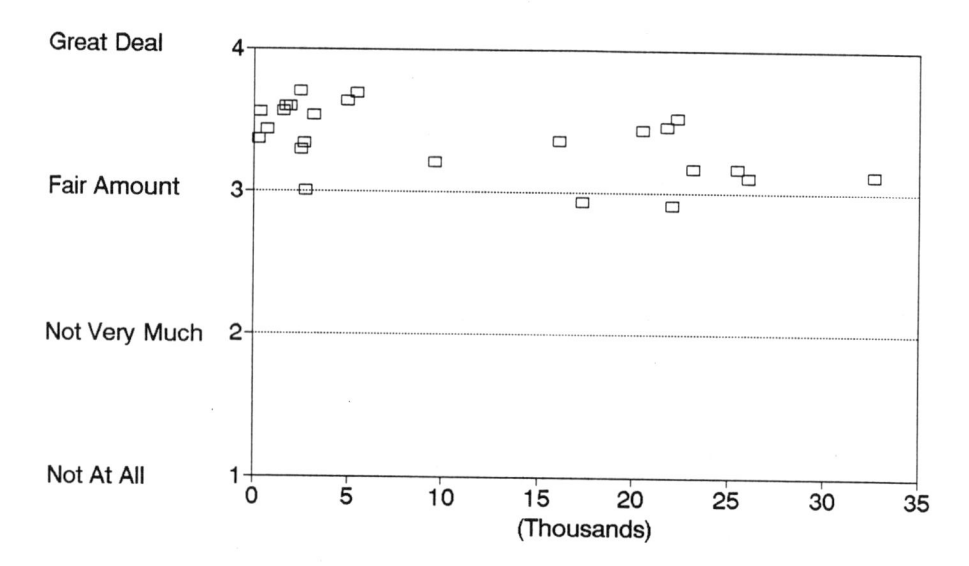

Figure 9. Perceived Health Effects of Environmental Problems in the Future By Per Capita GNP (1990 US Dollars)

Perceived Seriousness of Specific Environmental Problems

Besides asking respondents to rate the quality of the environment at various geographical levels, as noted above, we wanted to learn what types of environmental problems they viewed as most serious at each level. For the national level we used an open-ended question that asked respondents to volunteer what they considered to be the most important environmental problem (see Dunlap, et al., 1993a:18), but at the community and world levels we gave them lists of potential problems and asked them to rate the seriousness of each one.

For the community we used a list of six types of problems, and computed the mean of their ratings to create an overall indicator of **Average Perceived Seriousness of Six Community Environmental Problems**. As reported in Table 3, and shown in Figure 10, there is a significant negative relationship (r = -.52, p<.01) between national affluence and the overall perceived seriousness of environmental problems (e.g., air and water pollution) at the community level. Residents of the poorer nations are significantly more likely to see their local community as suffering from various forms of environmental degradation, just as they were more likely to rate the quality of their community environment as lower (as noted previously).

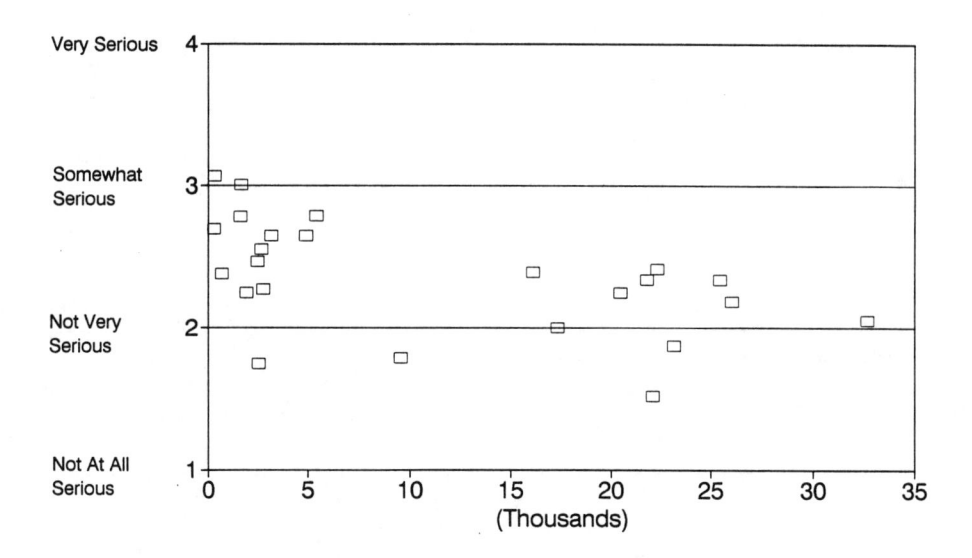

Figure 10. Average Perceived Seriousness of Six Community Environmental Problems By Per Capita GNP (1990 US Dollars)

We also computed the mean of respondents' ratings of seven worldwide environmental problems, which overall are rated as much more serious than are the community-level problems, to create a measure of **Average Perceived Seriousness of Seven World Environmental Problems**. Unlike the previously noted ratings of the quality of the world environment, the relationship is not the reverse of that for perceived seriousness of community environmental problems. Indeed, there is no relationship (r = .05, n.s.) between national affluence and the average perceived seriousness of specific environmental problems at the world level, as shown in Figure 11, even though residents of the poorer nations are more likely to respond that they 'don't know enough ... to judge' these problems (Dunlap, et al., 1993a)--a response treated as missing data for this variable. However, we should note that this is the one variable for which the results differ notably depending on whether one uses GNP/capita or log of GNP/capita, as in the latter case the correlation is more sizable (r = .30, n.s.) even though it still fails to reach significance. The at-best weak relationship between national affluence and citizens' perceptions of problems such as global warming and ozone depletion as serious will come as a surprise to those committed to conventional wisdom.

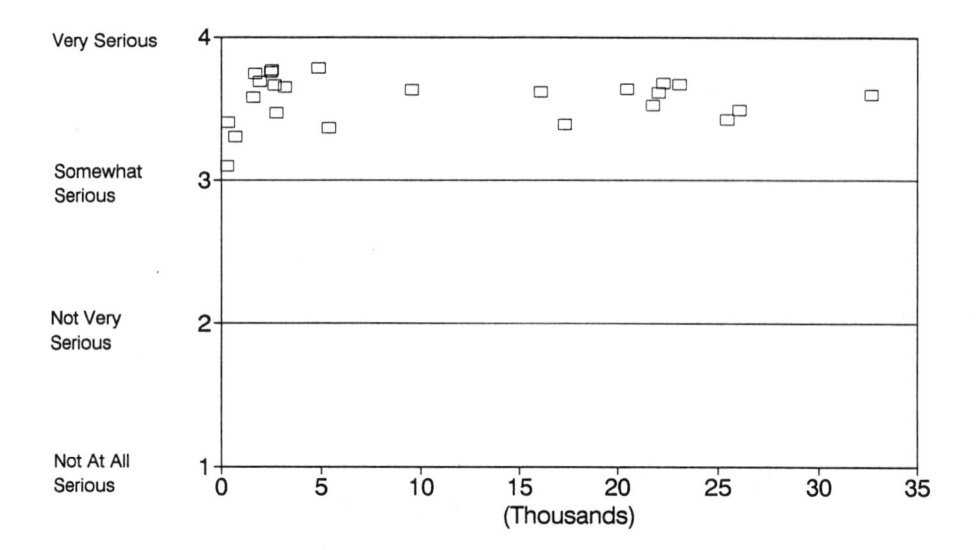

Figure 11. Average Perceived Seriousness of Seven World Environmental Problems By Per Capita GNP (1990 US Dollars)

Support for Environmental Protection Measures

In addition to inquiring about how serious respondents see environmental problems and how personally concerned they are about them, we wanted to examine levels of support for specific environmental protection measures that various nations might take. We included a list of six rather diverse measures in the HOP, and asked respondents whether they favored or opposed each one (the diversity of the measures likely accounts for the relatively low alphas for this variable obtained in some of the nations). To obtain an indicator of overall support for environmental protection, we computed the mean of the ratings given to these items. The resultant correlation between **Average Level of Support for Six Environmental Protection Measures** and per capita GNP provides one of the most striking contradictions to conventional wisdom, as it is strongly negative ($r = -.79$, $p < .001$). As shown in Figure 12, even though citizens in all 24 nations tend to favor these environmental protection measures, residents of the low income countries are clearly more likely to do so than are residents of high income countries. This result strongly reinforces the previously reported negative relationships, and indicates that residents of poorer nations not only tend to see environmental problems as more serious, but are also more supportive of efforts to ameliorate them, than are their counterparts in the wealthier nations.

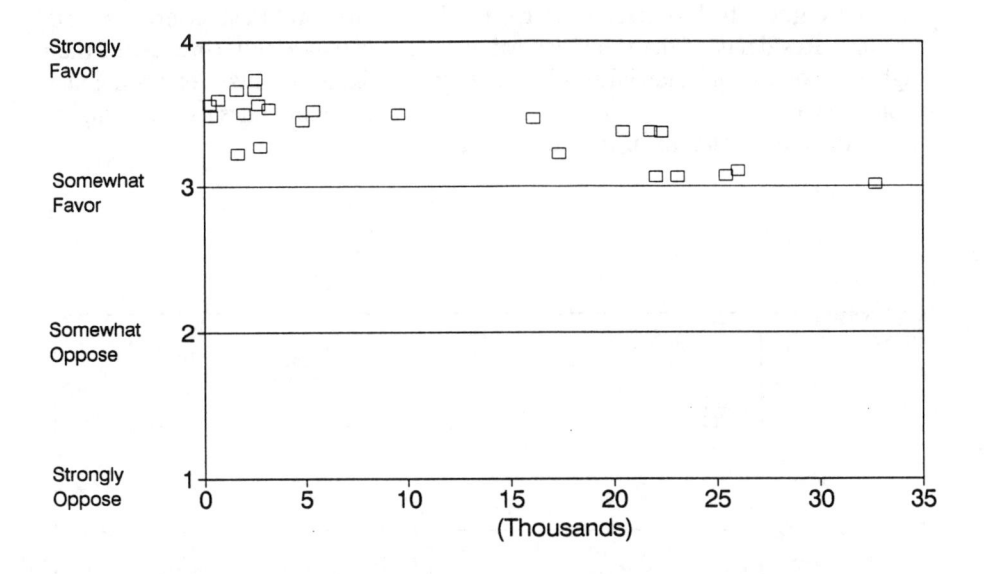

Figure 12. Average Level of Support for Six Environmental Protection Measures By Per Capita GNP (1990 US Dollars)

Environmental Protection and Economics

Despite the increasing emphasis being placed on the goal of sustainable development, or the achievement of economic growth without environmental deterioration, environmental protection is still widely viewed as conflicting with economic growth. Although aware of the potential shortcomings of questions that put the environment and economy in opposition, we nonetheless included a couple of items that forced respondents to make trade-offs between the two because they have been so widely used in the U.S.(Dunlap and Scarce, 1991). We did this reluctantly because we feared that such items were inherently biased against residents of low income nations, for even if the latter were as concerned about environmental quality as are their counterparts in the wealthier nations they might--of necessity--be less willing (and able) to make economic sacrifices.

The first item posed the environment vs. economy tradeoff at the societal level by asking respondents to choose between giving priority to environmental protection or to economic growth. The second item focused on the more personal level by asking respondents if they would, 'Be willing to pay higher prices so that industry could better protect the environment?' Not surprisingly, the correlations are significantly positive between per capita GNP and both **Preferred Trade-Off**

Between Environmental Protection and Economic Growth (r = .55, p < .01) and **Willingness to Pay Higher Prices for Environmental Protection** (r = .50, p < .05). Residents of the wealthier nations are understandably (in view of their higher incomes and standards of living) more likely to favor environmental protection over economic growth and to express a willingness to pay higher prices than are residents of the poorer nations.

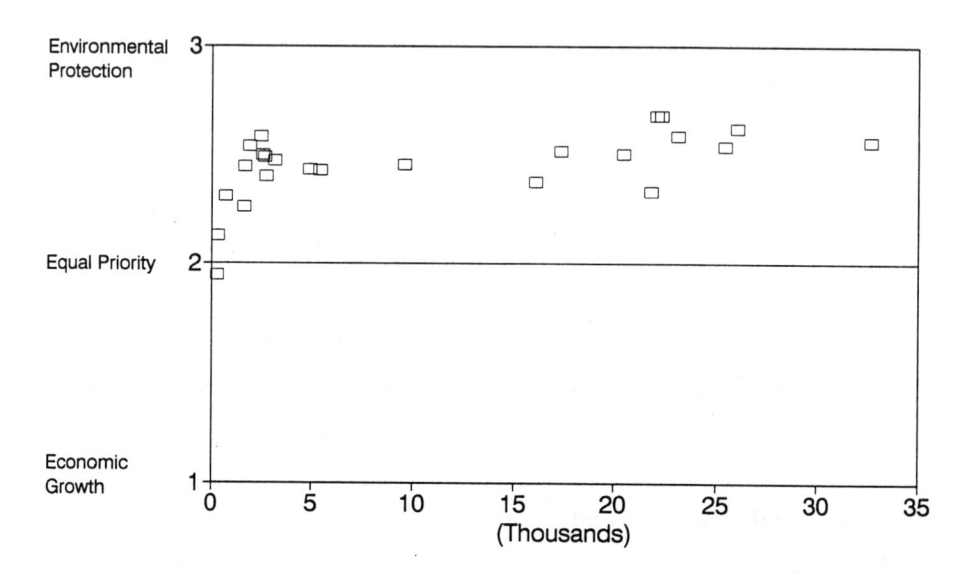

Figure 13. Preferred Trade-Off Between Environmental Protection and Economic Growth By Per Capita GNP (1990 US Dollars)

A couple of important features about responses to these two items, however, are revealed in Figures 13 and 14. First, citizens in a large majority of the 24 nations are more likely to emphasize environmental protection over economic growth or higher prices. Specifically, in all but one country (Nigeria) citizens overall give higher priority to environmental protection than to economic growth, and in only three are residents overall more likely to be unwilling than willing to pay higher prices (Nigeria, Philippines and, interestingly, Japan). Second, in both cases the significant correlation appears to be heavily influenced by results for a couple of the very poorest nations. In other words, although we find the expected positive relationships between national affluence and willingness to make economic tradeoffs on behalf of the environment, we should not lose sight of the fact that priority is still given to environmental protection in a majority of the low income countries[6].

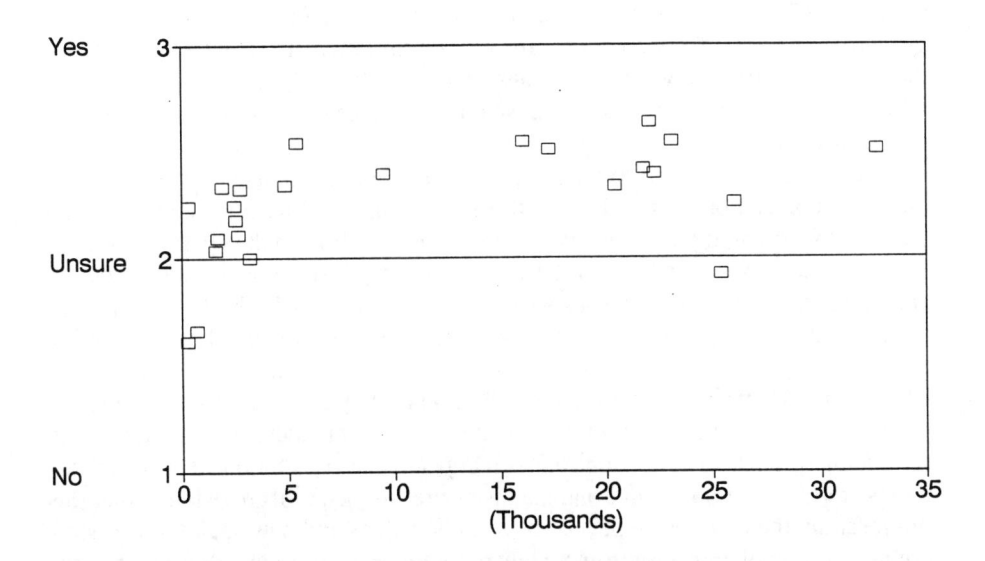

Figure 14. Willingness to Pay Higher Prices for Environmental Protection By Per Capita GNP (1990 US Dollars)

4. CONCLUSIONS

If it were true that generalized public concern about environmental quality is far more prevalent in the wealthy, industrialized nations than in poorer nations--as conventional wisdom has long suggested--then all of the correlations reported in Table 3 should be positive (or at least all but the one for perceived seriousness of environment relative to other national problems). However, only four of the environmental attitude variables are positively and significantly related to per capita GNP, while seven are negatively and significantly correlated with national affluence (three are not significantly correlated). Two of the positive correlations are for variables measuring environment-economic tradeoffs, measures for which residents of poorer nations are inherently disadvantaged, and a third is for the perceived seriousness of environmental problems relative to other national problems such as hunger and homelessness (which are surely more pressing in poor nations). The only other variable positively correlated with per capita GNP is rating of the quality of the world's environment. In contrast, the seven variables which are negatively correlated with GNP/capita cover a wide range of areas, including personal concern about environmental problems, perceived quality of community and national environment, perceived health effects from environmental problems (now and in the future), perceived seriousness of community environmental problems, and support for environmental protection measures. Clearly then, the preponderance of evidence contradicts the widespread view that citizens of poor nations are less environmentally concerned than are their counterparts in wealthy nations. Our results not only challenge 'lay' wisdom, but also conventional social science analyses of environmentalism. The idea that environmental quality is a luxury affordable only by those who have enough economic security to pursue quality of life goals is inconsistent with the observed correlations, as well as with the overall high levels of environmental concern found among residents of the low-income nations in the HOP. (While we likely would have found lower levels of concern in nations facing desperate situations like Somalia, countries such as Nigeria, India and the Philippines are nonetheless poor by any reasonable standard.) Our results are thus compatible with growing evidence of strong, grass-roots' environmentalism in much of the Third World (Durning, 1989; Finger, 1992), and suggest that such activism likely reflects broad-based concern for environmental quality in these nations rather than the anomalous acts of small minorities. It appears that conventional social science perspectives on global environmentalism are in need of revision. Theories premised on the emergence of post-materialist values and new social movements are useful in explaining environmentalism within most wealthy, Western nations (although Pierce, et al., 1987 question the applicability of the post-materialist explanation to Japan), but seem inappropriate for explaining the emergence of grass-roots environmentalism and widespread public concern for environmental quality in poorer nations. In part this may stem from the fact that environmental degradation is increasingly seen, especially in poor nations, not as a post-materialist quality-of-life issue but as a basic threat to human survival (as suggested by Ladd, 1982). In other words, environmental quality seems to be moving from a 'higher-order' value to a 'lower-order' need in Maslowian terms (see Dunlap, et al, 1983).

Future research on citizen concern about the state of the environment might benefit from paying more attention to direct experience with local environmental degradation (which will likely be more visible in poor nations) as well as to growing awareness of global environmental threats to human welfare (such as ozone depletion). It seems probable that awareness of the threats posed by environmental degradation at local, national and global levels--stimulated and reinforced by activists, scientists and media as well as some degree of personal observation--are beginning to reinforce one another. In some cases this may stimulate a homocentric concern with the welfare of one's self, family and descendants, but in other cases it may lead to a more ecological perspective that recognizes that human welfare is inextricably related to that of the environment (Oates, 1989; Stern and Dietz, 1994). Regardless of the nature and sources of public concern about environmental quality (which will no doubt vary across as well as within nations), the findings of the Health of the Planet survey should come as welcome news to those eager to move toward a more sustainable future. The old assumption that non-industrialized nations will not worry about environmental protection until they have achieved economic growth is incorrect. Citizens of these nations are clearly concerned about environmental quality; the question thus becomes whether or not the wealthy, highly industrialized nations (which have contributed disproportionately to world-wide environmental degradation) will assist them in translating their concern into effective action. The obstacles are many, but the rewards would be great.

REFERENCES

Baumol, W. J. and W.E. Oates. (1979). *Economics, Environmental Policy, and the Quality of Life*. Englewood Cliffs, NJ: Prentice Hall.

Beckerman, W. (1974). *Two Cheers for the Affluent Society*. New York: St. Martin's Press.

Brechin, S. R. and W. Kempton. (1994). 'Global Environmentalism: A Challenge to the Postmaterialism Thesis?' *Social Science Quarterly* 75:245-269.

Buttel, F. H. (1992). 'Environmentalization: Origins, Processes, and Implications for Rural Social Change.' *Rural Sociology* 57:1-27.

Dietz, T., L. Kalof and R.S. Frey. (1991). 'On the Utility of Robust and Resampling Procedures.' *Rural Sociology* 56:461-74.

Dunlap, R. E. (1995). 'Public Opinion and Environmental Policy.' In: J. P. Lester (ed.), *Environmental Politics and Policy* 2nd ed. Durham, NC: Duke University Press, pp. 63-113.

Dunlap, R. E., G.H. Gallup Jr. and A.M. Gallup. (1993a). *Health of the Planet*. Princeton, NJ: George H. Gallup International Institute.

Dunlap, R. E., G.H. Gallup Jr. and A.M. Gallup. (1993b). 'Of Global Concern: Results of the Health of the Planet Survey.' *Environment* 35 (November):6-15, 33-39.

Dunlap, R. E., J. K. Grieneeks and M. Rokeach. (1983). 'Human Values and Pro-environmental Behavior.' In: W. D. Conn (ed.), *Energy and Material Resources: Attitudes, Values and Public Policy*. Boulder, CO: Westview Press, pp. 145-168.

Dunlap, R. E. and R. Scarce. (1991). 'The Polls--poll Trends: Environmental Problems and Protection.' *Public Opinion Quarterly* 55:651-72.

Durning, A. (1989). 'Mobilizing at the Grassroots.' In: L. R. Brown, et al. (eds.), *State of the World 1989*. New York: Norton, pp. 154-73.

Elmer-Dewitt, P. (1992). 'Rich vs. Poor: Summit to Save the Earth.' *Time* (June 1):42-58.

Ester, P. and F. Van der Meer. (1982). 'Determinants of Individual Environmental Behaviour: an Outline of a Behavioural Model and Some Research Findings.' *Netherlands' Journal of Sociology* 18:57-94.

Finger, M. (1992). 'The Changing Green Movement--a Clarification. *Research in Social Movements, Conflicts and Change* Supplement 2:229-46.

Founex Report. (1972). 'Environment and Development.' *International Conciliation* No. 586:1-84.

Haas, P. M., M.A. Levy and E.A. Parson. (1992). 'Appraising the Earth Summit: How Should we Judge UNCED's Success?' *Environment* 34 (October):6-11, 26-33.

Inglehart, R. (1977). *The Silent Revolution*. Princeton, NJ: Princeton University Press.

Inglehart, R. (1990). *Culture Shift in Advanced Industrial Society*. Princeton, NJ: Princeton University Press.

Ladd, E. C. (1982). 'Clearing the Air: Public Opinion and Public Policy on the Environment.' *Public Opinion* 5 (February/March):16-20.

Leff, H. L. (1978). *Experience, Environment, and Human Potentials*. New York: Oxford.

Louis Harris and Associates (1989). *Public and Leadership Attitudes to the Environment*. New York: Louis Harris and Associates.

Murch, A. W. (1971). 'Public Concern for Environmental Pollution.' *Public Opinion Quarterly* 35:102-108.

Oates, D. (1989). *Earth rising: Ecological Belief in an Age of Science*. Corvallis: Oregon State University Press.

Pierce, J. C., N.P. Lovrich Jr., T. Tsurutani and T. Abe. (1987). 'Vanguards and Rearguards in Environmental Politics: A Comparison of Activists in Japan and the United States.' *Comparative Political Studies* 18:419-447.

Stern, P. C. and T. Dietz (1994). 'The Value Basis of Environmental Concern.' *Journal of Social Issues* 50:65-84.

Van Liere, K. D. and R.E. Dunlap. (1981). 'Environmental Concern: Does it Make a Difference How It's Measured?' *Environment and Behavior* 13:651-76.

World Bank. (1992). *World Development Report 1992*. Washington, D.C.: World Bank.

NOTES

1. Thanks are extended to George H. Gallup, Jr., Chairman of the Gallup International Institute, and Alec M. Gallup, Co-Chairman of the Gallup Organization, for making the data available, and to the National Science Foundation for a grant to support detailed analyses of the data set.

2. Basically Inglehart's (1977; 1990) theory of post-materialism represents a synthesis of Maslow's hierarchy of needs, Mannheim's theory of generational socialization, and Rokeach's value theory (Brechin and Kempton, 1994; Dunlap, et al., 1983).

3. The most comprehensive effort prior to the Gallup survey was a 16 nation survey conducted by Louis Harris and Associates (1989) for the United Nations Environment Program in 1988 and 1989. However, in most of the nations included in that study the samples were quite small (ranging from 300 to 600) and, in the non-industrialized nations, typically limited to residents of large urban areas, making it problematic to generalize results to the national populations.

4. Per capita gross domestic product is also frequently used, but we employ GNP/capita because we have an estimate of this figure for Russia by the Institute of Sociology of the Russian Academy of Sciences, and neither figure is provided for Russia by the World Bank (1992).

5. We will eventually employ more robust techniques such as bootstrapping and robust regression that are less sensitive to the violations of statistical assumptions common to cross-national data analyses (see, e.g. Dietz, Kalof and Frey, 1991).

6. It should also be noted that Brechin and Kempton (1994) use results from the Louis Harris multi-national survey noted above to compute the relationship between per capita GNP and two forms of "willingness to pay" for environmental protection-- higher taxes as well as volunteering two hours per week of community work on behalf of the environment. They found a modest, but insignificant, positive correlation ($r = .21$, $p<.44$) between national affluence and willingness to pay higher taxes but a strong and highly significant negative correlation ($r = -.78$, $p<.001$) between affluence and willingness to volunteer time. In other words, in the Harris survey residents of poor nations were a bit less willing to pay higher taxes, but far more willing to volunteer their time, on behalf of environmental protection. Brechin and Kempton conclude "that the observed reluctance to pay by [residents of] some of the poorest countries reflects extreme economic hardship, not a lack of environmental values." This conclusion is also compatible with the overall pattern of findings from the HOP survey.

Appendix: Country scores and Standard deviation

	USA	Canada	Brazil	Mexico	Russia	Portugal	Chile	Denmark	Finland	Germany	Hungary	India
Perceived Seriousness of Environmental Issues in Nation	3.44 (0.64)	3.46 (0.66)	3.25 (0.87)	3.56 (0.70)	3.59 (0.55)	3.36 (0.78)	3.46 (0.68)	3.01 (0.76)	2.92 (0.73)	3.63 (0.59)	3.43 (0.72)	3.40 (0.73)
Perceived Seriousness of Environment Relative to Other National Problems	-0.08 (0.62)	0.41 (0.68)	-0.22 (0.84)	0.19 (0.61)	0.24 (0.59)	0.07 (0.64)	0.05 (0.69)	0.62 (0.76)	0.45 (0.75)	0.61 (0.63)	0.24 (0.67)	-0.04 (0.72)
Personel Concern about Environmental Problems	3.23 (0.73)	3.26 (0.66)	3.27 (0.93)	3.29 (0.86)	3.19 (0.85)	3.37 (0.68)	2.99 (0.81)	2.62 (0.76)	2.78 (0.75)	2.73 (0.80)	3.10 (0.76)	3.07 (0.85)
Perceived Quality of Nation's Environment	2.55 (0.72)	2.26 (0.64)	2.67 (0.89)	2.76 (0.84)	3.33 (0.65)	2.36 (0.75)	2.86 (0.73)	2.06 (0.63)	2.09 (0.43)	2.47 (0.70)	2.91 (0.66)	2.66 (0.89)
Perceived Quality of Community Environment	2.16 (0.76)	2.00 (0.70)	2.49 (0.93)	2.20 (0.88)	3.01 (0.86)	2.18 (0.79)	2.41 (0.82)	1.76 (0.68)	2.02 (0.57)	2.12 (0.69)	2.61 (0.76)	2.51 (0.95)
Perceived Quality of World's Environment	2.92 (0.77)	3.07 (0.68)	3.05 (0.86)	3.14 (0.80)	3.09 (0.66)	3.23 (0.81)	3.34 (0.68)	3.39 (0.60)	2.88 (0.68)	3.24 (0.74)	3.03 (0.63)	2.71 (0.90)
Perceived Health Effects of Environmental Problems at Present	2.90 (0.93)	2.59 (0.90)	2.61 (1.15)	3.05 (1.11)	3.44 (0.69)	3.00 (0.93)	2.71 (0.92)	1.63 (0.80)	2.04 (0.72)	2.88 (0.83)	2.64 (0.84)	3.03 (0.90)
Perceived Health Effects of Environmental Problems in the Past	2.43 (0.96)	2.04 (0.85)	1.76 (0.96)	1.85 (0.97)	3.01 (0.89)	2.29 (1.01)	1.70 (0.80)	1.49 (0.74)	1.78 (0.65)	2.69 (0.87)	2.13 (0.81)	2.38 (0.86)
Perceived Health Effects of Environmental Problems in the Future	3.46 (0.72)	3.43 (0.69)	3.34 (0.90)	3.70 (0.65)	3.53 (0.69)	3.63 (0.61)	3.59 (0.61)	2.91 (0.80)	3.11 (0.66)	3.52 (0.71)	3.01 (0.90)	3.55 (0.69)

	USA	Canada	Brazil	Mexico	Russia	Portugal	Chile	Denmark	Finland	Germany	Hungary	India
Average Perceived Seriousness of Six Community Environmental Problems	14.02 (4.35)	13.46 (4.07)	15.30 (5.15)	14.77 (5.42)	15.87 (4.36)	15.85 (5.10)	13.47 (4.73)	9.09 (3.36)	13.12 (4.42)	14.46 (4.36)	13.64 (3.88)	18.37 (4.06)
Average Perceived Seriousness of Seven World Environmental Problems	24.66 (3.34)	25.44 (2.67)	25.60 (2.88)	26.26 (2.93)	25.52 (2.84)	26.43 (2.17)	25.78 (2.68)	25.23 (2.77)	24.40 (3.21)	25.69 (2.87)	24.28 (3.11)	23.80 (3.11)
Average Level of Support for Six Environmental Protection Measures	20.25 (3.05)	20.25 (2.48)	21.38 (2.74)	21.98 (2.52)	21.21 (2.11)	20.69 (2.51)	21.02 (2.49)	18.38 (2.78)	18.63 (2.53)	20.21 (2.73)	19.65 (2.54)	20.89 (2.55)
Preferred Trade-Off between Environmental Protection and Economic Growth	2.33 (0.86)	2.50 (0.77)	2.49 (0.83)	2.58 (0.72)	2.47 (0.67)	2.43 (0.68)	2.54 (0.69)	2.68 (0.64)	2.62 (0.65)	2.68 (0.58)	2.40 (0.71)	2.13 (0.85)
Willingness to Pay Higher Prices for Environmental Protection	2.42 (0.85)	2.34 (0.88)	2.11 (0.98)	2.24 (0.93)	2.00 (0.89)	2.34 (0.89)	2.33 (0.93)	2.63 (0.73)	2.26 (0.85)	2.39 (0.81)	2.32 (0.76)	2.24 (0.92)

	Japan	Netherlands	Nigeria	Philippines	Poland	Korea	Turkey	Denmark	Finland	Norway	Ireland	Britain
Perceived Seriousness of Environmental Issues in Nation	3.34 (0.69)	3.09 (0.69)	3.19 (0.88)	3.13 (0.81)	3.63 (0.57)	3.61 (0.62)	3.51 (0.71)	3.22 (0.84)	3.56 (0.65)	3.28 (0.70)	3.08 (0.80)	3.19 (0.73)
Perceived Seriousness of Environment Relative to Other National Problems	0.76 (0.66)	0.75 (0.75)	-0.47 (0.84)	-0.14 (0.74)	0.20 (0.56)	0.64 (0.62)	0.03 (0.63)	-0.07 (0.72)	0.65 (0.75)	0.52 (0.69)	0.02 (0.73)	0.05 (0.76)
Personel Concern about Environmental Problems	2.94 (0.75)	2.87 (0.69)	3.63 (0.72)	3.49 (0.62)	2.12 (0.72)	3.02 (0.68)	2.28 (0.99)	3.15 (0.83)	2.41 (0.88)	2.95 (0.69)	2.90 (0.79)	3.06 (0.74)
Perceived Quality of Nation's Environment	2.65 (0.76)	2.51 (0.58)	2.49 (0.97)	2.61 (0.80)	3.33 (0.73)	3.01 (0.79)	2.42 (0.92)	2.36 (0.74)	2.19 (0.62)	1.96 (0.55)	1.99 (0.64)	2.44 (0.70)
Perceived Quality of Community Environment	2.29 (0.69)	2.20 (0.62)	2.37 (0.95)	2.21 (0.75)	3.02 (0.87)	2.70 (0.90)	2.48 (1.03)	2.22 (0.78)	2.07 (0.66)	1.78 (0.65)	1.78 (0.70)	2.23 (0.78)
Perveived Quality of World's Environment	3.27 (0.63)	3.18 (0.61)	2.35 (0.99)	2.75 (0.78)	3.06 (0.69)	3.08 (0.70)	2.58 (0.95)	3.20 (0.83)	3.23 (0.68)	3.21 (0.65)	3.03 (0.75)	3.09 (0.72)
Perceived Health Effects of Environmental Problems at Present	2.23 (0.72)	2.35 (0.65)	3.24 (1.05)	2.94 (0.85)	3.30 (0.82)	3.03 (0.79)	2.90 (1.09)	2.29 (1.19)	2.06 (0.84)	1.95 (0.94)	2.35 (1.03)	2.56 (0.94)
Perceived Health Effects of Environmental Problems in the Past	2.01 (0.70)	2.01 (0.66)	2.50 (1.15)	2.48 (0.89)	2.87 (0.97)	1.77 (0.65)	2.03 (1.04)	1.73 (1.01)	1.92 (0.77)	1.70 (0.89)	1.89 (0.88)	2.09 (0.87)
Perceived Health Effects of Environmental Problems in the Future	3.17 (0.69)	2.93 (0.66)	3.36 (0.95)	3.43 (0.80)	3.60 (0.67)	3.69 (0.65)	3.56 (0.79)	3.29 (0.94)	3.12 (0.82)	3.17 (0.75)	3.21 (0.87)	3.35 (0.76)

	Japan	Netherlands	Nigeria	Philippines	Poland	Korea	Turkey	Uruguay	Switzerland	Norway	Ireland	Britain
Average Perceived Seriousness of Six Community Environmental Problems	14.01 (3.86)	11.99 (3.23)	16.13 (4.46)	14.28 (4.66)	18.03 (4.05)	16.70 (4.21)	16.67 (5.15)	10.47 (4.06)	12.30 (4.15)	11.23 (4.85)	10.69 (4.61)	14.36 (4.69)
Average Perceived Seriousness of Seven World Environmental Problems	23.95 (3.42)	23.73 (2.94)	21.69 (5.42)	23.09 (4.09)	26.19 (2.61)	23.55 (3.16)	25.02 (3.32)	26.36 (2.66)	25.14 (2.99)	25.64 (2.71)	25.39 (2.89)	25.29 (2.90)
Average Level of Support for Six Environmental Protection Measures	18.43 (2.21)	19.36 (2.52)	21.37 (2.47)	21.56 (2.46)	19.37 (4.36)	21.12 (2.25)	21.96 (2.14)	22.42 (1.93)	18.07 (3.19)	18.39 (3.03)	20.98 (2.33)	20.79 (2.40)
Preferred Trade-Off between Environmental Protection and Economic Growth	2.54 (0.58)	2.51 (0.62)	1.95 (0.81)	2.31 (0.88)	2.44 (0.73)	2.43 (0.81)	2.26 (0.74)	2.49 (0.74)	2.56 (0.62)	2.59 (0.73)	2.45 (0.80)	2.38 (0.78)
Willingness to Pay Higher Prices for Environmental Protection	1.92 (0.84)	2.51 (0.74)	1.61 (0.90)	1.66 (0.91)	2.09 (0.95)	2.54 (0.77)	2.03 (0.93)	2.18 (0.94)	2.50 (0.82)	2.54 (0.79)	2.39 (0.81)	2.54 (0.75)

9. Individual Change and Stability in Environmental Concern: The Netherlands 1985-1990

Peter Ester
Loek Halman
Brigitte Seuren

Abstract[1]

Public opinion polls have repeatedly shown that advanced modern societies are characterized by enduring and rather stable levels of high environmental concern. Some suggest that this high level of environmental concern is a result of a paradigm shift from an industrial towards a post-industrial worldview. Whether or not the adoption of such a post-industrial worldview is indeed inducive to a higher environmental concern at the individual level is investigated analyzing panel data from the Dutch SOCON study in 1985 and 1990.
An important result is that the overall stable trend at the aggregated level seems to mask attitudinal turbulency at the individual level. Approximately half of the Dutch citizens switched their position with respect to environmental issues! Further, our analyses challenge the idea that environmental concern is strongly dependent upon a post-industrial worldview. In fact the results of our analyses indicate mixed support for this hypothesis. A majority of Dutch citizens appears to be on the midway between a rejection of the old industrial worldview and the adoption of a new post-industrial worldview.

1. INTRODUCTION

Environmental problems are unmistakably a salient and lasting issue on the public and political agenda in the Netherlands at the end of the twentieth century. In view of the seriousness of these problems, there is ample reason for this high level of attention. Environmental concern has accelerated to a degree, that according to Nelissen (1990:15) Dutch society is presently characterized by what he calls 'green hyperactivity'. Environmental issues have increasingly become the subject of ordinary, day-to-day communication between citizens. Public institutions at all levels (national, regional and municipal) are highly concerned with the framing and implementation of active environmental policies. The mass media continue to draw public attention to environmental problems, and countless conferences are devoted to this issue. In industrial circles, the idea of corporate environmental care systems has become a hot item. In the Netherlands hundreds of active environmental groups have been established, and the national environmental movement is a major interlocutor and a strong solicitor for the formulating and evaluating of environmental policy. Last, but certainly not the least, Dutch citizens are actively concerned with the environment in terms of attitudes and behaviors. However, their behaviors do not run entirely parallel to their attitudes as will be seen later in this chapter.

Mobilization for societal problems, such as the environmental one, appears to be cyclical. Downs (1972) has called this 'the issue-attention cycle'. Nelissen (1990:14-16) has reconstructed this cycle for the environmental issue. He distinguishes eight stages (see also: Ester and van der Meer, 1979a; Ester and Mandemaker, 1993):

1. *Signalizing the environmental issue.* In this stage concerned scientists try to draw public attention to the dramatic changes in the relationship between humans and their environment. The best known examples were Rachel Carson with her book *Silent Spring* (1962) and, above all, Meadows and his associates with the *Report to the Club of Rome* (1972).
2. *Recognition of the problems.* In this second stage environmental action groups arise, who have adopted the concern of the intellectual vanguard. After having experienced several environmental incidents, they demand governmental action in order to prevent more environmental deterioration.
3. *Preliminary measures.* In this stage concern for the environmental issue is spreading rapidly, until broad segments of society have reached a fairly high level of concern. The first attempts at issuing environmental policy are made.
4. *The 'legalistic' approach.* This stage is characterized by the implementation of a series of legal environmental measures by the government, urged by the environmental movement. The industrial circles remain more or less outside this process, or even oppose it.
5. *Preventive approach.* In this stage it becomes increasingly clear, that legal instruments alone are not sufficient to prevent environmental pollution or deterioration. Preventive actions are necessary to establish this. The kind of policy characteristic in this stage is 'resource policy', in which target groups are being postulated.
6. *Broadening the initiatives.* In this stage there is active involvement of hitherto rather passive actors, such as trade and industry. In dialogue with the authorities, they want to arrive at clear and feasible appointments concerning the reduction of emissions. The authorities, on their turn, reflect the effectiveness of their instruments in use: legal measures, impositions, subsidies, education, etc.
7. *Environmental wave.* This seventh stage shows all kinds of distinct organizations, groups and individuals initiating activities to prevent further environmental pollution and deterioration. All such activities are

summarized by the term 'green hyperactivity', and according to Nelissen (1990) Dutch society at the beginning of the nineties seems to be in this phase.
8. *The environment becoming commonplace.* In this final stage, green hyperactivity is no longer seen as deviant behavior, but has become ordinary practice. Environmentally friendly thinking and acting has become the rule and has been internalized. In short, environmentally conscious activities have become routine.

This model is of course a somewhat sketchy reconstruction of the developments that have occurred in the past few decades concerning environmental issues. It nevertheless provides a clear picture of the broadening of the social mobilization for the environment[2]. 'Sustainable development' has become the cornerstone of environmental policy, both nationally and internationally. Particularly the report *Our Common Future* (WCED, 1987) has been a great source of inspiration (see also the Dutch *National Environmental Policy Plan*, NMP, 1989; and its successors). Although the concept itself is rather vague (Opschoor and van der Ploeg, 1990), it tries to unite, on a worldwide scale, the purposes of both developmental and environmental policy. Sustainable development reflects a development in which the interests of coming generations are very explicitly taken into account. This implies the abandonment of old values and norms and the encouragement of more environmentally friendly lifestyles. In view of the severity of the environmental problems, this process has to proceed boldly, in which clear boundaries have to be set on the human use of the environment. The Dutch Committee for Long-Term Environmental Policy (CLTM, 1990) argues that radical changes are necessary, may be even an 'environmental revolution'. The adoption and diffusion of a sustainable and environmentally friendly lifestyle presupposes a change of fundamental values. Values characteristic in an industrial-capitalistic society (like hedonism, consumerism, unlimited economic growth etc.) have to give way to values which are consistent with a sustainable development (like environmental concern, taking account of future generations, modified and restrained production and consumption, attention to the quality of life instead of quantity and for human scale, being aware of resource shortages, etc.). The adoption and diffusion of more environmentally friendly lifestyles requires a high and continuous level of environmental concern in the different segments of society. This is the main theme of this chapter: what can be said about the environmental concern of the Dutch population and which trends can be traced in its environmental concern? More specifically we are interested in the stability of environmental concern over time at the *individual* level. The reason for this is directly linked to the crucial notion of sustainable development as it presupposes the continuity of environmental concern at the level of individual citizens.

In this chapter we will analyze whether environmental attitudes are characterized by stability over time, and whether attitudinal stability is related to sociodemograhics and to more general social and political beliefs. This is a rather novel approach to the study of environmental concern as most studies are cross-sectional (cf. Ester, 1979; Van der Meer, 1981; Ester and Nelissen, 1992; Nelissen and Scheepers, 1992). Panel studies of the development of environmental attitudes over the individual lifecourse are almost absent (cf. Ester, Seuren and Nelissen, 1994). This omission is an unfortunate one, both from a policy and sociological point of view. As indicated above the diffusion and adoption of sustainable development assumes continuity and persistence of environmental concern at the level of the individual citizen and consumer. A development towards sustainable lifestyles will only take place if environmental concern is deeply rooted in the individuals' mental framework. The necessary routinalization of pro-environmental behavior is based on the postulation that environmental concern is widely and continuously internalized. Various studies have indeed indicated that environmental concern is widespread and has become less of a typical 'middle-class' issue but is to be found across social-demographic borders (Ester and Mandemaker, 1994). Recently the cross-national *Health of the Planet Survey* (Dunlap et al., 1993) indicated that the concern for the environment is a global one trespassing traditional divisions such as North/South and East/West (see also the chapter in this book by Dunlap and Mertig). The main conclusion from this large scale international study is the existence of a 'worldwide citizen awareness that our planet is indeed in poor health, and great concern internationally for its future wellbeing. The results not only document widespread citizen awareness and concern, but highlight the existence of a more worldwide consensus about environmental problems than is widely assumed' (Dunlap et al., 1993:38). The finding that environmental awareness and concern is widely shared is a major precondition for the acceptance of sustainable lifestyles. However, what is needed most is continuity of pro-environmental attitudes of citizens expressing high awareness and concern. Monitoring these attitudes presupposes repeated panel research and this is exactly the method that is being used in this chapter.

Reaching the final stage of the environmental issue-attention cycle is not merely a matter of stable environmental concern. As will be argued in the next section of this chapter, the likelihood of stability of pro-environmental attitudes is enhanced if these attitudes are structurally backed up by and are part of a more encompassing worldview that stresses more general basic social values that are consonant with the principle of sustainable development. The switch from environmentally unfriendly lifestyles towards sustainable lifestyles is not just a question of higher and stable environmental concern, but first and foremost it requires a fundamental shift in the current 'social paradigm' characteristic of modern society, i.e. a shift from a 'technological paradigm' towards an 'intrinsically ecological' one. The issue of how stability of pro-environmental attitudes is empirically related to basic social values underlying this paradigm shift, is the second objective of this chapter. Within this context the question will also be addressed whether this paradigm shift in basic social values reflects generational differences.

The structure of this chapter is as follows. Section 2 outlines the main hypotheses underlying this chapter and section 3 introduces the data that are used for testing these hypotheses and summarizes the main measurement instruments. Sections 4 through 6 describe the main outcomes which are discussed in section 7.

2. HYPOTHESES

The main focus of this chapter is on trends in environmental concern and its determinants at the *individual* level. We are particularly interested in how prevailing worldviews affect (in)stability of individual environmental attitudes. Our approach to this question has been inspired by a recent study of Olsen, Lodwick and Dunlap (1992). According to these authors a qualitative change in the value system of Western industrialized society is taking place at present. This change is from a 'prevailing industrial worldview' towards an 'emerging post-industrial worldview'. Both value systems are characterized by radically different social paradigms: the 'technological social paradigm' and the 'ecological social paradigm', respectively. Emphasis on material achievements, economic efficiency, instrumental rationality, 'big is beautiful', quantitative ends, cultural homogeneity and an orientation towards the present are values creating the heart of the 'industrial worldview'. Personal development, social effectiveness, valuebound rationality, human scale, qualitative ends, cultural heterogeneity and orientations towards the future are values constructing the nucleus of the 'post-industrial worldview'. According to the research results of Olsen et al. a majority of Americans would already subscribe to the ecological paradigm. The basic hypothesis that will be tested in this chapter is that *the more people adhere to the post-industrial worldview, the higher and more stable their environmental concern over time will be.*

Thus, the level and stability of attitudes towards environmental problems are supposed to be directly linked to individuals' support for basic values underlying the ecological paradigm. The more these basic values have been internalized, the higher and more stable a persons's environmental concern will be. In this sense environmentalism is seen as an intrinsic part of a broader value system which reinforces strength and continuity of environmental concern.

The shift from an industrial worldview towards a post-industrial one parallels the shift Inglehart (1990) observes from materialist towards post-materialist values in advanced industrial societies. According to Inglehart the unprecedented levels of economic growth and development, the emergence of the welfare state, the rising levels of education and the vast expansion of mass media have gradually but consistently transformed basic beliefs that make up the culture of modern society. The realization of economic and physical security that is characteristic for the welfare state induced a shift from values stressing material well-being towards values emphasizing quality of life. Postmaterialist values accentuate goals such as self-actualization, individual autonomy and freedom, social belonging, aesthetic fulfillment, intellectual challenges, and cultural and economic skills. As younger generations have been socialized in relatively secure economic and physical circumstances Inglehart holds (and shows) that younger generations in particular support postmaterialist values: 'the young emphasize postmaterialist goals to a far greater extent than do the old, and cohort analysis indicates that this reflects generational change far more than it does aging effects' (Inglehart, 1990:103). As these goals can be seen as illustrative for a post-industrial worldview as discussed above[3], it can be deducted that younger generations more so than older generations will adhere to the post-industrial ecological paradigm. Following the first hypothesis this in turn implies that *stability of environmentalism is particularly present among younger generations*. As environmental concern is an intrinsic feature of postmaterialism and the post-industrial paradigm, and as this paradigm is pre-eminently supported by younger generations, these generations will show higher stability in pro-environmental attitudes than older generations as it is structural part of a more encompassing worldview.

3. DATA AND MEASUREMENT INSTRUMENTS

Although interest in environmental issues on the part of the social sciences is rather recent, both in the Netherlands and abroad[4], a well grounded tradition has developed in social scientific research on environmental concern and behavior in the past two decades in the Netherlands (Nelissen and Schreurs, 1975; Ester, 1979a, 1979b; Van der Meer, 1981; Ester, 1984; Nelissen et al., 1987, 1988; Ester and Seuren, 1992; Nelissen and Scheepers, 1992; Ester et al., 1993). This chapter positions itself explicitly into that tradition, both conceptually and methodologically.

The data to be analyzed in this chapter are part of a large scale longitudinal survey project on social and cultural developments in the Netherlands. The data have been collected in 1985 and in 1990. The project, well known as the SOCON project, dates back to 1979. In this year a nationwide survey called *Secularization and depillarisation in the Netherlands* was conducted by Felling, Peters and Schreuder (1986). The aim of this survey was to investigate, among other things, the influence of church involvement and religious beliefs on nonreligious attitudes and behaviors, e.g., fundamental values, political views, cultural and economic conservatism in Dutch society. In order to investigate attitudinal and behavioral changes a repeated survey was carried out in 1985. The scope of the study was enlarged by including attitudes and behaviors on a wider range of social issues, among which environmentalism (Felling, et al. 1987). In 1990 a third wave was conducted covering the same

domains as in 1985. These longitudinal cross-sectional data enable comparisons over time on an aggregate (i.e. societal) level (Ester and Halman, 1994).
However, a number of researchers involved in the SOCON project were not only interested in aggregated societal changes but also in developments at the individual level. Therefore it was decided that a major part of the 1990 study was designed as a panel. Respondents interviewed in 1985 were again approached in 1990 to be interviewed on the same issues as five years before. From the original sample of 3003 respondents, 1794 persons appeared to be available for the 1990 wave; 1344 of them were actually approached. A total of 311 respondents seemed to be included erroneous, so the net result was 1033 respondents who were asked to be interviewed again. Of this sample 350 persons refused to collaborate, yielding a total of 683 respondents that actually cooperated. As a consequence of the specific design of the 1985 questionnaire 333 respondents have answered the questions on environmentalism (for extensive and more detailed information see Felling, Peters and Schreuder, 1987; Felling, Peters and Scheepers, 1992).

The measurement of environmentalism in the SOCON project distinguishes three components: the attitude towards the environment, offering willingness for the natural environment, and finally action willingness. These three components or dimensions of environmentalism have been used in sociological and psychological environmental research in the Netherlands as early as 1975 (Nelissen and Schreurs, 1975). For each component several items are formulated, which are mentioned in Table 1.

Table 1 Overview of items on environmentalism and mean scores (1 = agree entirely; 5 = don't agree at all) in 1985 and 1990

Attitude towards the natural environment:	1985	1990
- Protecting rare plants and animals is an unnecessary luxury	4.04	4.04
- For me it is not necessary to protect unspoiled nature at any cost	3.79	3.97

Offering willingness for the natural environment:		
- I am willing to give up something to get a more beautiful environment	2.47	2.19
- I would be willing to pay higher prices for products if that would mean less industrial pollution	2.28	2.12
- It is a good thing that the government wants to fight waterpollution, but it must not cost me a penny	3.62	3.85

Action willingness for the natural environment:		
- I would take part in a demonstration against more and more industrialisation	3.78	3.60
- I am willing to propagate a strict reduction of motorised traffic	3.19	3.05
- To join a demonstration against felling trees is not like me	2.66	2.76
- If there was a protest meeting against the start of a polluting factory in the neighboorhood, you could expect to find me there	2.82	2.50
- I shall join in with protest actions, that try to do anything against acid rain	2.84	2.78

A look at the distributions of the answers to the questions demonstrates that the appreciation of the natural environment is widespread among Dutch people: three out of every four Dutch respondents agrees with both statements. More variation is found on the items concerning offering- and action willingness. However, a three dimensional pattern appears to be valid at both measurement points in time. Factor analyses performed on the 1985 and 1990 data separately yielded evidence for such a pattern, and the similarity of the pattern was substantiated by applying LISREL analysis.

Factor scores were estimated for each factor separately using the factor score coefficients based on the 1985 data. This approach enables to investigate changes over time in the respondents position on these dimensions. To trace such individual changes, it is necessary that the same metric for the estimation of scores on the latent concept is applied at both points in time. The main reasons to derive this metric from the 1985 sample are:
1. the 1985 sample has proven to represent the Dutch population (to a large extent);

2. the 1985 sample is relatively large compared to the panel sample of 1990, and the larger the sample the smaller the margins of unreliability.

After having estimated the individual factor scores in 1985 and 1990, difference scores were calculated by subtracting the 1985 score from the 1990 score. This score represents the individual changes in the three environmental attitudes (Felling, Peters and Scheepers, 1992:238). These difference scores enable us to further explore the characteristics of those who have developed higher environmental concern in the 1985-1990 period and those that indicated lower levels of environmentalism. What distinguishes them from one another and what are the main differences with those who remained rather stable in their stands towards environmental issues? Apart from standard socio-demographic characteristics, such as age, gender, education, income, region, it has been investigated whether or not groups of respondents showing increased, stable or decreased environmental concern differ in their basic worldviews, or more precisely whether these groups adhere to a respectively emerging post-industrial worldview or to a prevailing industrial worldview.

We have argued that stable and high levels of environmental concern can be seen as a crucial part of what is called the post-industrial worldview which is the core concept of the ecological paradigm. This worldview refers to a complex of societal values and beliefs emphasizing individual self-realization and concerns for collective goods. The main distinctive characteristics can be summarized as follows:

industrial worldview	**post-industrial worldview**
material achievement	personal development
economic efficiency	social effectiveness
instrumental rationality	valuebound rationality
big is beautiful	human scale
quantitative ends	qualitative ends
cultural homogeneity	cultural heterogeneity
orientation towards present times	future orientation

According to the ecological paradigm a strong environmental concern is accompanied by a less traditional orientation in the domain of religion, social criticism in politics and a predominant post-materialist orientation. Such orientations are measured in the SOCON project by various indicators which were combined in validated scales (see Felling, Peters and Schepers, 1992 for an overview).
The traditional Christian worldview consists of the Christian interpretation of an ultimate and transcendent reality and a similar interpretation of the meaning of life, death and suffering. A secular worldview means that such issues are interpreted from a world-directed and immanent perspective. Social criticism includes items such as related to the importance of the reduction of income

differences and the diminishing of power distances. A short version of Inglehart's post-materialism scale was used covering items about self-development and personal autonomy on the one hand and items on maintaining order and authority on the other hand.

Post-industrial worldviews are further reflected in progressive individual and societal developments as manifested in an emancipated view on women and the resistance against the reduction of civil liberties. In contrast to the adherents of the ecological paradigm, it is assumed that people who are not, or to a lesser extent, concerned about the environment, will be less politically and socially active, and that they will be more strongly directed towards their own comfort. They are less interested in politics in general and less inclined to take part in various political activities like conventional and unconventional protest activities. Their geo-political scope is limited and mainly focussed on the local community and not on the broader society.

Indicators of these religious, political, and social orientations are widely available in the SOCON data in the form of scales. These scales are either factor scores or sum scores and they have proven to be valid and reliable in numerous studies (Felling, Peters and Scheepers, 1992; Ester and Halman, 1994).

4. INDIVIDUAL CHANGES IN ENVIRONMENTAL CONCERN

This section addresses the issue of individual change or stability of environmental concern in the 1985-1990 period. In Table 2 these individual changes are shown after having trichotimized the scores on the three sub-dimensions[5]. The crosstabulations not only yield the proportions of respondents who have changed their attitudes, but also indicate in which direction they have changed. Did they become more or less environmentally concerned?

Table 2 Individual changes in environmentalism (in %) (1985-1990)

	attitude towards nature	offering willingness	action willingness
decrease	22	36	27
stable	50	47	53
increase	30	17	21

From Table 2 it becomes clear that about half of the Dutch the population did not alter its environmental attitudes. The highest stability exists with regard to action willingness (53%), the lowest with respect to offering willingness (47%). However, Table 2 also demonstrates that half of the Dutch has indeed changed! And that is even the more remarkable given the observation that at the aggregate level hardly any evidence for changes in environmentalism in Dutch society was obtained (Ester et al. 1994:198). These aggregate level data apparently do obscure rather massive changes at the individual level. Obviously, aggregated data on societal stability of environmental concern do not reflect what is really going on at the level of individual environmental attitudes. The minor changes at the aggregate level seem to be the result of both negative and positive changes at the individual level. These changes appear to compensate each other, particularly in case of the attitude towards the natural environment and action willingness. The proportions positive changes are almost equal to the proportions of negative changes. Only with regard to offering willingness a slight decrease can be noted. The proportion of respondents who have become less willing to make sacrifices for the environment surpasses the proportion of cases who are increasingly willing to contribute. These findings clearly illustrate that overall stability in environmental concern does not imply stability at the individual level. More precisely: observed macro stability of environmental attitudes in Dutch society is in fact the result of an almost perfect compensation of positive and negative micro changes in environmental concern. Thus, aggregated stability in environmental attitudes masks rather substantial changes at the individual level.

The main changes as described in Table 2 do not reveal which of the various environmental opinions have changed most. Therefore in Table 3 the individual changes in the separate items on environmentalism are presented.

Table 3 Individual changes in individual items on environmentalism (in %) (1985-1990)

	757	758	766	767	769	770	771	772	774	776
-4	1.5	-	-	-	-	-	-	-	-	.3
-3	1.8	2.1	.6	1.2	1.5	1.0	.6	2.5	1.0	1.3
-2	4.2	4.9	3.4	1.8	3.4	2.6	6.7	7.2	4.5	3.3
-1	15.8	14.6	9.3	11.2	8.2	16.1	18.2	14.4	13.7	23.8
0	51.2	43.2	52.6	61.2	57.6	43.8	42.0	46.6	43.3	38.4
1	20.0	25.8	26.0	21.2	23.2	27.0	23.9	18.8	20.1	25.4
2	4.2	6.4	7.1	2.1	5.2	8.6	8.0	7.8	14.0	5.5
3	.6	2.7	.6	.9	.6	1.0	.6	2.8	3.5	2.0
4	.6	.3	.3	.3	.3	-	-	-	-	-
mean	-.03	.17	.25	.11	.17	.23	.08	.07	.33	.07
st dev	1.12	1.15	.93	.84	.93	1.00	1.06	1.19	1.17	1.09

Interpretation of variables:

757 Protecting rare plants and animals is an unnecessary luxury
758 For me it is not necessary to protect unspoiled nature at any cost
766 I am willing to give up something to get a more beautiful environment
767 I would be willing to pay higher prices for products if that would mean less industrial pollution
769 It is a good thing that the government wants to fight waterpollution, but it must not cost me a penny
770 I would take part in a demonstration against more and more industrialisation
771 Even I am willing to propagate a strict reduction of motorised traffic
772 To join a demonstration against felling trees is not like me
774 If there was a protest meeting against the start of a polluting factory in the neighboorhood, you could expect to find me there
776 I shall join in with protest actions, that try to do anything against acid rain

Table 3 illustrates that at the individual level action willingness has increased in the 1985-1990 period. The proportions of Dutch citizens who have become more willing to take part in various activities in favor of the environment surpasses the number of people who have become less eager to get actively involved. Particularly the willingness to attend a protest meeting against the construction of a polluting factory in one's neighborhood has increased. The changes in the items on action willingness are more obvious than in case of offering willingness and the general attitude towards the environment. This last result will be due to the widespread environmental concern already found in 1985. An increase could hardly be expected to have taken place as a result of a ceiling effect.

The individual changes in the general attitude towards the environment, offering willingness and action willingness, can not be attributed to specific demographic groups within Dutch society. Regression analyses in which the three difference scores are the dependent and socio-demographics the independent variables provided no evidence for the existence for such groups. This appears also in Table 4 where the mean difference scores are presented for various demographic groups.

Table 4 Mean difference scores on attitude towards nature, offering willingness and action willingness for various socio-demographic groups

	attitude towards nature	offering willingness	action willingness
Age			
18-30	.02	.05	.13
31-45	.06	-.08	.10
46-60	-.02	-.13	.01
60+	-.13	.00	-.16
eta	.11	.06	.11
p	.26	.73	.34
Education			
elementary school	-.12	.12	-.04
lower vocational school	-.04	-.01	-.21
lower secondary school	.26	-.00	.33
secondary vocational school	-.07	-.14	-.04
o levels/a levels	-.04	-.18	-.17
college	-.08	-.14	.22
university*	-.07	-.01	.35
eta	.09	.09	.23
p	.89	.79	.03
* N = 8			
Income			
< 1500 guilders	-.30	-.04	-.31
> 1500 and ≤ 2500 guilders	-.04	.09	-.31
> 2500 and ≤ 3250 guilders	-.21	-.12	.07
> 3250 and ≤ 5000 guilders	-.04	-.18	.02
> 5000 guilders	.28	.01	.31
eta	.16	.11	.24
p	.09	.51	.00
Gender			
male	.01	-.19	.01
female	-.08	.07	..06
eta	.04	.12	.03
p	.47	.02	.62

Exceptions are education and income. The higher one's education and the higher one's income, the greater the willingness is to take part in various actions in favor of the environment. However, in case of education this relationship is not linear, for not only the highest educated people have changed in this direction, those having lower secondary schooling have changed similarly.

Regression analyses in which these three difference scores are regarded the dependent variables and the indicators of a post-industrial worldview the independent ones, yielded similar results as were found with regard to the socio-demographic background variables: they were disappointing. No evidence could be found that changes in environmental attitudes are a specific feature of certain groups in society according to their value orientations. Thus, adhering to a post-industrial worldview as such does not influence micro stability or change in environmental concern.

It also has been investigated whether or not the individual changes in environmental attitudes could be attributed to individual changes in post-industrial worldviews. In other words, it has been examined whether those who have embraced the post-industrial worldview have altered their environmental attitudes too. For that reason difference scores were calculated for each of the value orientations we have described before, and they were included in the regression analyses. The results of these analyses were, again, disappointing. Apparently, no evidence could be found that changes in environmental attitudes were due to, or better accompanied by, individual changes in basic value orientations. Hence, the conclusion is that individuals' changes in environmental outlook can not be attributed to individual's changes in worldview.

5. ENVIRONMENTAL SWITCHERS AND STABLE TYPES

In order to further examine the changes in environmental concern in Dutch society, a typology was constructed based on the dimensions 'offering willingness' and 'action willingness'. The 'general attitude towards the environment' is excluded from this typology because of the very skewed distribution. As before both attitudes have been trichotomized and then combined as explained in Table 5.

Table 5 Construction of the environmental typology

		action willingness		
		1	2	3
offering	1	1	1	2
willingness	2	1	2	3
	3	2	3	4

The nine logically possible classes have been reduced to four types which can be interpreted as follows:
1. the environmentally apathetics;
2. the environmentally unconcerned;
3. the environmentally concerned;
4. the environmentally friendly.

As demonstrated in Table 6 environmentally friendly attitudes have decreased in the 1985-1990 period which corroborates the results at aggregate level. At the aggregated level it was concluded that the environmentally friendly group diminished slightly, as was the case with the environmentally apathics. The group of concerned people decreased in number too, whereas a relatively sharp increase was found in the proportion of the environmentally unconcerned. At the individual level similar results were obtained. However, the differences are less pronounced at the individual level compared to the differences found at the aggregated level. The group of environmental friendly people has decreased, whereas the other types increased slightly.

Table 6 Shifts in environmental types (1985-1990)

year	apathetics	unconcerned	concerned	friendly
1985	28	27	21	24
1990	32	29	23	15

A total of 55% of the environmentally apathics in 1985 were still indifferent in 1990. The stability of the other environmental types was more limited. About one third of its members did not alter their views, and thus stayed in the same category. Most of the switchers move one position, either to the negative side or to the positive side. Some 39% of the environmentally friendly in 1985 did not belong to this group anymore in 1990.

In Table 7 the individual shifts are indicated. The triangle at the left and bottom includes the cases who have altered their position in an environmentally negative direction. Those who are in the upper right triangle have developed more environmentally friendly attitudes.

Table 7 Environmental types (%) in 1985 and 1990

		1990 apathetics	unconcerned	concerned	friendly
	apathetics	15	9	3	1
1985	unconcerned	11	9	5	2
	concerned	4	8	7	3
	friendly	2	4	9	9

The number of people who have become less concerned about the environment in the 1985-1990 period exceeds the number of those who have become more concerned (38% versus 23%). The rest (39%) did not alter their position. Consequently, in terms of our typology, a downward 'trend' in environmental concern at the individual level is observed.

6. CHARACTERISTICS OF ENVIRONMENTAL TYPES

What can be said about the socio-demographic characteristics of those who have changed in either an environmentally positive or negative direction and those who have remained stable? And what about the assumed emergence of a new ecological paradigm among those with stable environmental attitudes and those who have altered their environmental attitudes in a positive direction? Which conclusions hold for those who have become less environmentally concerned? Do they support the old industrial paradigm?

In Table 8 an overview of the main characteristics of the three environmental groups is presented.

Table 8 Background characteristics for negative switchers, stable people and positive switchers (mean scores and %)

	negative	stable	positive
Age	46.41	46.03	45.03

eta = .04

Income	%	%	%
< 1500 guilders	44	31	25
> 1500 and ≤ 2500 guilders	33	47	19
> 2500 and ≤ 3250 guilders	35	48	17
> 3250 and ≤ 5000 guilders	40	39	21
> 5000 guilders	38	30	32

Education	%	%	%
elementary school	38	38	24
lower vocational school	32	47	21
lower secondary school	45	32	23
secondary vocational school	37	44	19
o levels/a levels	50	32	18
college	36	36	28
university*	28	29	43

*N = 8

Gender	%	%	%
male	37	42	21
female	39	36	25

Region	%	%	%
north	35	44	21
east	50	31	19
west	41	37	23
south	26	46	28

Church involvement	%	%	%
2nd generation unchurched	43	40	17
1st generation unchurched	42	37	21
former church members	40	29	31
marginal church members	42	40	19
modal church members	29	51	21
core church members	36	36	28

It will be clear that in general the differences between the three environmental types are rather modest, and far from significant. However, the higher income groups and the higher educated groups are slightly more likely to have changed their environmental outlook in a positive direction compared to people with lower incomes and lower levels of education. Furthermore, it can be concluded that a negative change occurred particularly in the eastern and western regions of the Netherlands, whereas people in southern regions are more likely to have changed their environmental view in a more positive direction. Former church members, as well as core church members, appear to be more concerned about the environment than five years ago. The fact that such a large proportion of core church members has changed in a positive direction contradicts the expectation that it is particularly people adhering to secular worldviews that will support the new ecological paradigm! This is also reflected in the finding that those who have changed their environmental attitudes in a positive direction or stabilized their attitudes are more supportive of the Christian worldview than those who have changed their environmental outlook in a negative direction. This pattern becomes clear from table 9.

Table 9 Indicators of post industrial world view for negative switchers, stable people and positive switchers (mean scores)

	negative	stable	positive	eta
christian worldview	476	502	493	.12
world-directed worldview	508	500	520	.09
traditional achievement	493	503	513	.08
social criticism	505	508	524	.08
economic conservatism	504	508	511	.03
cultural conservatism	460	478	458	.10
rejection civil liberties	1.66	1.98	1.68	.09
rejection intervention life and death	1.04	1.34	1.14	.11
traditional view on women	481	469	458	.08
localism	488	486	458	.13
postmaterialism	6.11	5.82	5.75	.07
political interest	517	537	540	.11
conventional pol. participation	2.56	2.66	2.56	.03
unconventional pol. participation	2.67	2.48	3.00	.11
political apathy	4.97	4.66	5.08	.10
political alienation	475	479	455	.10

However, table 9 also reveals that positive switchers are also more in favor of a world directed worldview, which is more in line with basic characteristics of a post-industrial worldview. The findings provide mixed support for the hypothesis that the more people adhere to the post-industrial worldview, the higher and more stable their environmental concern is. There is evidence that positive switchers and people with stable environmental attitudes are somewhat more stressing achievement orientations, social criticism, emancipated views on women, show higher levels of political interest and unconvential political participation and lower levels of political alienation and localism compared to negative switchers. However, contrary to the hypothesis negative switchers are more likely to be less conservative in economic terms and they reject to a lesser degree the intervention in matters of life and death, they are more supportive of postmaterialist views and show lower levels of political apathy. In some cases the attitudes of people with stable environmental attitudes (e.g. cultural conservatism, convential political participation) are noteworthy. Thus, there is no unambigious proof that the emerging new ecological paradigm is adopted *in toto* by people who became more environmentally conscious. At present this adoption process is rather diffuse and somewhat ambivalent. It seems as if Dutch society is somewhere between an industrial and a post-industrial worldview combining elements of both. This cultural transition stage apparently struggles with the adoption of new and the abandoning of old values.

The second hypothesis stated that stability of environmentalism is particularly present among younger generations as they are most likely to adhere to the new ecological paradigm and the post-industrial worldview underlying this paradigm. Results are shown in Table 10.

Table 10   Environmental attitude change in age groups (in %)

age	negative	stable	positive
23-27	35	35	29
28-32	27	46	27
33-37	37	37	26
38-42	40	38	22
43-47	35	46	19
48-52	44	40	16
53-57	54	21	25
58-62	22	56	22
63-67	38	31	31
68-75	40	45	15

It is immediately clear from Table 10 that stability of environmentalism is not a unique feature of the youngest age group. Some 35% of this age group holds stable environmental attitudes, whereas 56% of the 58-62 years of age did not alter their positions. The group of 53-57 years of age has most drastically changed its environmental attitudes in a negative direction. The figures in Table 10 shows a very capricious pattern which is not easily to interpret. In short, the hypothesis suggesting that stability (and increase) of environmental concern is a specific feature of young people has obviously to be refuted.

7. CONCLUSIONS

Successive public opinion polls in recent years have repeatedly shown that advanced modern societies are characterized by enduring and rather stable levels of high environmental concern. This conclusion is substantiated by the other empirical chapters in this part of this volume. Data analyzed in this chapter indicate that also in the Netherlands the public appears to be deeply concerned about environmental issues. A vast majority of Dutch citizens shows a positive attitude towards the environment and shows dito levels of offering willingness. Action willingness to preserve the environment - though less widely adopted - appears a well-rooted behavioral dimension of environmental concern. Our analyzes covering the 1985-1990 period reveal fairly stable macro levels of environmental concern in Dutch society. However, this overall stable trend at the aggregated level masks attitudinal turbulency at the individual level. Our panel data showed that approximately half of the Dutch citizens switched their position with respect to environmental issues! Due to compensating effects substantial changes at the individual level, environmental concern at the macro level hardly changed in the 1985-1990 period. Minor changes at the aggregate level seem to be the result of both significant negative and positive changes at the individual level. This is a highly remarkable and notable result as it indicates that macro-analyzes of stable trends in environmental concern clearly mask momentous changes at the individual level. Thus, observed macro-stability of environmental concern may very well be the result of compensating decreases and increases in concern at the individual level. The message from these findings for mainstream public opinion research on environmental issues is clear: studies showing that the public's environmental attitudes in Western societies are both highly positive and stable should not be presented as evidence for micro-stability in environmental attitudes. The public's attitudes towards environmental issues are in fact quite dynamic and far from unchanging.

The leading hypothesis tested in this chapter suggested that the adoption of a post-industrial worldview - rooted in an emerging new ecological paradigm, will induce higher environmental concern at the individual level. This post-industrial worldview is characterized by stressing personal development, social effectiveness, human scale, postmaterialism, valuebound rationality, qualitative ends and a distinctive future orientation. Our analyses do not substantiate the assumed intercorrelation between environmental concern and the adherence to a post-industrial worldview. In fact the results of our analyses indicate mixed support for this hypothesis. Though there exists some evidence that in Dutch society in the early nineties people did adopt elements of this new ecological paradigm, it must be recognized that this paradigm as such is not convincingly supported. It seems more likely that the majority of Dutch citizens is on the midway between a rejection of the old industrial worldview and the adoption of a new post-industrial worldview. No clear generational or age cleavages were observed in this respect. Apparently people are experimenting with this new ecological paradigm but the industrial paradigm has not yet been abandoned. Thus, although there is some support for an emerging paradigm shift, it seems as if Dutch citizens are still struggling with the question to what degree this new paradigm needs to be adopted. Such a situation is highly typical of a cultural transition phase such as the one we are now witnessing.

References

Carson, R. (1962). *The Silent Spring*. Houghton Mifflin.
CLTM. (1990). *Het Milieu: Denkbeelden voor de 21ste Eeuw*. Zeist: Kerckebosch.
CLTM. (1994). *Towards a Sustainable Future*. Dordrecht: Martinus Nijhoff Publishers.
Downs, A. (1972). 'Up and Down with Ecology - the Isssue Attention Cycle'. *The Public Interest* 2:38-50.
Dunlap, R., G.H. Gallup and A.M. Gallup. (1993). *The Health of the Planet Survey*. A George H. Gallup Memorial Survey. Princeton: Gallup International Institute.
Ester, P. (1979a). *Milieubesef en Milieugedrag*. Een Sociologisch Onderzoek naar Attitudes en Gedragingen van de Nederlandse Bevolking met Betrekking tot het Milieuvraagstuk. Amsterdam: Instituut voor Milieuvraagstukken, Vrije Universiteit.
Ester, P. (ed.). (1979b). *Sociale Aspecten van het Milieuvraagstuk*. Assen: Van Gorcum.
Ester, P. (1984). *Consumer Behavior and Energy Conservation*. A Policy-oriented Field Experimental Study on the Effectiveness of Behavioral Interventions Promoting Residential Energy Conservation. Erasmus Universiteit Rotterdam.
Ester, P., L. Halman and B. Seuren. (1993). 'Environmental Concern and Offering Willingness in Europe and North America'. In P. Ester, L. Halman en R. de Moor (eds.), *The Individualizing Society*. Value Change in Europe and North America. Tilburg: Tilburg University press, p. 163-182.
Ester, P. and T. Mandemaker. (1994). 'Socialization of Environmental Policy Objectives: Tools for Environmental Marketing'. In: CLTM, *Towards a Sustainable Future*. Dordrecht: Martinus Nijhoff Publishers.
Ester, P., B. Seuren and N. Nelissen. (1994). 'Een "Groene Golf" in Nederland? Milieubesef en Milieugedrag van Nederlanders'. In: P. Ester and L. Halman (eds.), *De Cultuur van de Verzorgingsstaat*. Een Sociologisch Onderzoek naar Waardenoriëntaties in Nederland. Tilburg: Tilburg University Press, p. 193-214.
Halman, L., J. Maas and N. Nelissen. (1992). *Milieugedrag van Consumenten*. Tilburg: IVA.
Inglehart, R. (1990). *Culture Shift in Advanced Industrial Society*. Princeton: Princeton University Press.
McGuire, W.J. (1985). 'Attitudes and Attitude Change'. In: E. Aronson and G. Lindzey (eds.), *Handbook of Social Psychology*. Vol.II: Special Fields and Applications. New York: Random House.
Meadows, D. e.a. (1972). *Rapport van de Club van Rome*. Utrecht/Antwerpen: Het Spectrum.
Meer, Van der. F. (1981). *Attitude en Milieugedrag*. Leiden (dissertatie).
Nelissen, N.J.M., R. Perenboom, P. Peters and V. Peters. (1987). *De Neder-*

landers en hun Milieu. Een Onderzoek naar het Milieubesef en het Milieugedrag van Vroeger en Nu. Zeist: Kerckebosch.

Nelissen, N.J.M. (1992). *Afscheid van de Vervuilende Samenleving?* Zeist: Kerckebosch (Inaugurele rede Katholieke Universiteit Brabant).

Nelissen, N.J.M. and P. Scheepers. (1992). 'Ecological Consciousness and Behavior Examined'. *Sociale Wetenschappen* 35:64-81.

NIPO. (1991). *Tekstrapport Nulmeting Milieu Gedrags Monitor*. Den Haag: Rijksvoorlichtingsdienst.

Nationaal Milieubeleidsplan: Kiezen of Verliezen. (NMP) (1989). 's-Gravenhage: SDU.

Nationaal Milieubeleidsplan-plus. (NMP-plus) (1990). 's-Gravenhage: SDU.

Olsen, M.E., D.G. Lodwick and R.E. Dunlap. (1992). *Viewing the World Ecologically*. Boulder, Col.: Westview.

Opschoor, J.B. and S.W.F van der Ploeg. (1990). 'Duurzaamheid en Kwaliteit: Hoofddoelstellingen van Milieubeleid'. In: CLTM (1990).

Schreurs, L. and N.J.M. Nelissen. (1975). *Het Meten van Milieubesef*. Den Haag, Maastricht, Nijmegen: Raad der Europese Gemeenten, Ministerie van CRM, Sociologisch Instituut Universiteit Nijmegen.

Tellegen, E. and M. Wolsink. (1992). *Milieu en Samenleving. Een Sociologische Inleiding*. Leiden/Antwerpen: Stenfert Kroese.

NOTES

1. Parts of this chapter have originally been published in Dutch: P. Ester, B. Seuren and N. Nelissen. (1994). 'Een "groene golf" in Nederland?'. In: P. Ester and L. Halman (eds.), *De Cultuur van de Verzorgingsstaat. Een Sociologisch Onderzoek naar Waardeoriëntaties in Nederland*. Tilburg: Tilburg University Press, p. 193-214. Furthermore the chapter is based on parts of Seuren's thesis which will appear in 1996 at Tilburg University Press.

2. It is possible for different parts of the entire environmental issue to be in different stages of the attention cycle. Recycling of materials, energy saving and separation of different kinds of household trash are in the final stages of the cycle, whereas attention for the climatic changes still has to make the larger part of the journey through the cycle, being at present roughly on the border passing from stage two to three.

3. In fact Inglehart is using the term 'worldview' himself to put postmaterialist values in a broader cultural perspective (Inglehart, 1990: 422).

4. There are several explanations for the rather late 'discovery' of the relationship between humans and their natural environment by social scientists, and above all by sociology. One of these explanations is, that the classic Durkheimian methodology of explaining social phenomena primarily by other social phenomena, has remained in use for a very long time.

5. The scores are factor scores (mean = 0, st dev. = 1) and are trichotomized as follows: low thru -.33 = 1; -.33 thru .33 = 2; .33 thru high = 3.

10. Farmers' Attitudes to Environmental Issues: an Australian Study

Alan W. Black
Ian Reeve

Abstract[1]

This chapter analyses survey data from a nation-wide sample of Australian farmers (N = 2044), showing how attitudes to various rural environmental issues vary across different agricultural industries and relate to social characteristics of farmers, as well as to characteristics of their farms. From a pool of 75 Likert-type items, eight orthogonal attitudinal factors or dimensions are identified, each with its own set of indicators. Non-attitudinal variables explain more than 10 per cent of the variance on each of the first four factors, but the particular combination of non-attitudinal variables is different for different factors. Likely reasons for these findings are discussed.

1. INTRODUCTION

Concerns about the environmental impacts of agriculture in Australia have received varying public attention since first European settlement. Until the 1940s, much of the land degradation that occurred was viewed more as proof of the strangeness and perverseness of the Antipodes, than as a threat to future prosperity (Powell, 1976). During the 1930s and 1940s, there was a growing awareness that soil erosion threatened to reduce the productivity of agriculture (see, for example, Jacks and Whyte, 1939). With the passage of soil conservation legislation in most States of Australia, State agencies mounted extension programs encouraging landholders to instal structural works to control runoff, to reduce the use of bare fallowing and to stock more conservatively. The environmental movement that arose in the 1960s directed its attention largely to wilderness preservation and urban pollution, leaving land degradation outside the ambit of its concerns. However, during the 1980s, agricultural and pastoral land degradation became increasingly to be seen as an environmental issue, leading to the publication of substantial studies such as those edited by Chisholm and Dumsday (1987) and Cameron and Elix (1991). The emergence of the sustainable development concept in the late 1980s (World Commission on Environment and Development, 1987) and its translation into ecologically sustainable development as a policy goal for agriculture in Australia (Commonwealth of Australia, 1990) has ensured that concerns about land degradation will stay within the ambit of environmental debate.

One expression of such concern during the past ten years in Australia has been the rapid growth of the landcare movement, composed mainly of local groups of rural landholders who aim to deal with problems of land degradation and to develop more sustainable farming systems and land-management practices (Campbell, 1991; Roberts, 1992; Black and Reeve, 1993). This movement was given a significant boost by the decision of the Australian government in 1989 to designate the decade beginning in 1990 as 'The Decade of Landcare' and to provide over $320 million for landcare and related tree planting and remnant vegetation conservation programs to the year 2000. Nevertheless, the voluntary nature of the landcare movement, pressures from an increasingly environmentally aware urban majority for reduced environmental impacts by agriculture, demographic change in rural Australia, growing scarcity of water and intact riverine ecosystems and increased health consciousness among consumers of agricultural produce are all factors that contribute to the increasing likelihood of the introduction of various policies designed to ensure that essential ecosystem functions are not degraded by agricultural activities. The success or failure of such policies will depend greatly upon their acceptability within the farming community, and therefore upon the extent to which policy development takes account of farmers' attitudes and orientations towards ecological issues. Although some studies have been made of rural/urban differences in environmental concern (e.g. Buttel and Flinn, 1974; Glenn and Alston, 1977; Tremblay and Dunlap, 1978; Lowe and Pinhey, 1982; Mohai and Twight, 1986; Freudenberg, 1991) and other studies have examined farmers' attitudes to particular farm practices which may affect the environment — such as the use of agricultural chemicals (e.g. Buttel *et al.*, 1981; Hoiberg and Bultena, 1981; Gillespie, 1987; Gillespie and Buttel, 1989) or the adoption of conservation technologies (for a review, see Black and Reeve, 1993) — there have been few *comprehensive* studies of farmers' attitudes to environmental issues directly associated with farming. The research reported in this chapter is intended to be a step in that direction. It is based on data from a 1991 mail survey of a nation-wide sample of Australian farmers (N=2044).

2. THEORETICAL CONSIDERATIONS

In analysing the data from a survey such as this, it is important to identify the factors which, on the basis of prior theorizing and research, one might expect to influence farmers' perspectives on rural environmental issues, including their attitudes to related public policy options. In some cases, different theoretical considerations would lead us to hypothesize different, even opposite, outcomes from the operation of a particular factor. In such cases, empirical research might help us to decide between these competing theoretical principles; but it is also possible that the competing tendencies neutralize one another. From the theoretical principle that peoples' economic interests tend to shape their attitudes to economically relevant issues, one would expect that farmers' attitudes would be influenced by the particular types of agricultural production in which they are engaged. For example, those involved in activities which are heavily dependent on the use of agricultural chemicals are likely to view with disfavour any policy which increases the cost of such chemicals or impedes their use. By contrast, those practising organic or alternative forms of farming are likely to favour policies which curb the use of agricultural chemicals. Even within conventional types of agriculture, some industries — such as grass-fed beef production or wool growing (industries known in Australia as grazing) — are often less dependent on agricultural chemicals than are others such as fruit growing, cotton growing and grain legume production. Attitudes toward the regulation of chemical use are likely to vary accordingly. This is not to say that attitudes can easily be predicted from the extent of usage. Those who depend heavily on the use of chemicals may also believe that manufacturers should be required, by law, to guarantee both the efficacy and the safety of the chemicals they produce. Consequently, the relationship between types of agricultural production and attitudes towards public policies affecting the use of agricultural chemicals could be quite complex. Depending on whose interests are affected, the same may be true of the relationship between types of production and attitudes towards other environmentally related issues.

Scale of farming operation, often measured by farm size, is another structural variable whose effects need to be considered. Other factors being equal, one would expect that the larger the farm the greater the likelihood that capital intensive methods of production will be used. Hence, one could hypothesize that farm size is positively associated with opposition among farmers to policies that would restrict the use of such methods. However, Hoiberg and Bultena (1981) found in Iowa a negative bivariate association between farm size and the opinion that there is too much government regulation pertaining to the use of pesticides and feed additives and to the safety of farm machinery. On the other hand, they found a positive bivariate association between farm size and disapproval of government-sponsored programs aimed at promoting soil conservation, but no significant association between farm size and attitudes to government regulations on solid waste disposal. Hoiberg and Bultena concluded that the variability in these bivariate relationships is due in part to differences in the forms of government involvement, which vary in their degree of compulsion and in cost-benefit ratio for the farmer. Form of land tenure and level of equity are further factors which may be relevant. Although it has often been hypothesized that owner-operators are more likely than short-term tenants to adopt conservation practices, empirical evidence does not always support this hypothesis (Earle, Rose and Brownlea 1979; Lee, 1980; Ervin and Ervin, 1982; Bultena and Hoiberg, 1983; Lee and Stewart, 1983; Ervin, 1986; Norris and Batie, 1987; Lynne, Shonkwiler and Rola, 1988). Again, some writers have hypothesized an inverse relationship between debt level and the adoption of conservation practices (Ervin and Ervin, 1982; Norris and Batie, 1987), while others have hypothesized a direct relationship but at a declining rate as debt load increases (Lynne,

Shonkwiler and Rola, 1988). Neither of these hypotheses has been consistently supported in these studies (see also Earle, Rose and Brownlea, 1979). Given this evidence, it remains to be seen whether form of land tenure and level of equity affect attitudes to other rural environmental issues. The influence of off-farm employment and off-farm income is also open to debate. On the one hand, off-farm income makes one less dependent on farm income and therefore less threatened by any decline in immediate productivity consequent upon greater restrictions in the use of particular forms of technology that are thought to be environmentally damaging. On the other hand, off-farm work reduces the time available for farm work and therefore makes one more likely to use chemical inputs and other potentially polluting technology rather than labour-intensive alternative methods (Pfeffer, 1992). These different tendencies are likely to interact in complex ways with the gender division of labour on farms and with other factors such as farm size and commodities produced. For example, farms on which the husband works off-farm and the wife does not are generally smaller than those in which the wife works off-farm and the husband does not (Coughenour and Swanson, 1983; Fassinger and Schwarzweller, 1984; Rosenfeld, 1985). Again, part-time farming is more common in some types of agricultural production such as beef cattle, horses and fruit than in others such as dairy cattle, field corn, soybeans, wheat and cotton (Wimberley, 1983; Buttel and Gillespie, 1984; Barlett, 1986; Simpson, Wilson and Young, 1988).

Turning to gender as an independent variable, writers such as McStay and Dunlap (1983), George and Southwell (1986) and Blocker and Eckberg (1989) have argued that women have generally been socialized into nurturant roles and that this translates *inter alia* into a desire to protect the environment, whereas men have generally been socialized into instrumental roles and tend therefore to adopt an exploitative attitude towards the environment. Consistent with these propositions, women have been found to be more concerned than men about nearby environmental hazards, especially health and safety hazards, such as those associated with nuclear power plants (Passino and Lounsbury, 1976; Brody, 1984; George and Southwell, 1986; Solomon, Tomaskovic-Devey and Risman, 1989), proposed coal-based industrial developments in a rural area (Stout-Wiegand and Trent, 1983), aerial spraying with pesticides (Hawkes *et al.*, 1984), toxic chemical wastes (Hamilton, 1985a), and local air and water pollution (Hamilton, 1985b; Blocker and Eckberg, 1989). However, the evidence of gender differences has been less clear-cut in studies of attitudes to more general environmental issues (i.e. to issues not specifically limited to one's neighbourhood or community), with some studies showing no significant differences between men and women, some showing women to have slightly higher levels of concern, and others showing men to be slightly more concerned than women (Blocker and Eckberg, 1989; Steger and Witt, 1989; Arcury and Christianson, 1990; Jones and Dunlap, 1992; Mohai, 1992). The theorizing and research reviewed in this paragraph have dealt with men and women in general, rather than with farm men and farm women in particular. The main conclusion to emerge is that differences in attitude between men and women are likely to be more evident on some environmental issues than on others. Women are likely to be more concerned than men about direct threats to health and safety, especially at the local level. As previous research appears to have neglected the influence of gender on farmers' attitudes towards environmental issues, the present chapter gives explicit attention to this issue.

In the population at large, age is generally the best sociodemographic predictor of environmental concern, with younger adults tending to be more environmentally conscious than older ones (Van Liere and Dunlap, 1980; Honnold, 1981; Mohai and Twight, 1987; Jones and Dunlap, 1992). Two main explanations have been suggested for this finding:

- The first explanation focuses on life-cycle factors, hypothesizing that the young are less integrated into the dominant social and economic order and that they are therefore more willing than their elders to accept environmentally beneficial changes to existing social institutions and patterns of behaviour. An implication of this hypothesis is that as they become older and more integrated into the dominant order, people become less willing to accept such potentially radical changes.

- The second explanation focuses on period or cohort effects, hypothesizing that more recent cohorts have, especially during their youth and young adulthood, become conscious of environmental and other social problems on a scale not experienced by earlier cohorts, and that this has left an indelible imprint on them. An implication of this hypothesis is that as they become older their concern for environmental issues will not necessarily diminish.

In an analysis of time series data from the General Social Survey, Honnold (1984) concluded that the lower level of environmental concern among older age groups is largely accounted for by period effects, but that a life-cycle effect is also evident among young adults. Although they did not use time series data, Mohai and Twight (1987) argued somewhat similarly that age differences in environmental concern are due largely to cohort effects.

Despite these findings about environmental concern within the general populace, the evidence on the relationship between age and farmers' concern about environmental issues has been rather mixed. Buttel *et al.* (1981) found that age and concern about pollution from agricultural chemicals were negatively associated among farmers in New York State but were not significantly associated among farmers in Michigan. Hoiberg and Bultena (1981) found in Iowa that age was not significantly associated with the opinion among farmers that there is too much government regulation pertaining to solid waste disposal and to animal feed additives; however, age had a positive bivariate association with disapproval of government-sponsored soil conservation programs and a negative bivariate association with the opinion that there is too much government regulation relating to pesticides and to the safety of farm machinery. On the other hand, Gillespie and Buttel (1989) found no significant association between age and degree of opposition among farmers in New York State to government regulation of agricultural chemicals and pharmaceuticals. These rather mixed results may be partly due to the fact that age interacts with other variables, some associated with life-cycle and some not. For example, younger farmers tend to be better educated and thus more aware both of environmental problems and of possible solutions. They may also be more energetic and thus more willing to deal with the difficulties of obtaining information and the complexities of monitoring associated with non-chemical methods of pest control. On the other hand, they may lack the financial resources needed to adopt some farming systems. This in turn could be related to their scale of operation, to how long they have been farming, to whether they have inherited or bought the farm or whether they are lessees, to their farm business structure, to their level of debt and to whether they have off-farm employment or income. Along with age, each of these factors will be considered in the present chapter. Educational attainment has been found not to be significantly associated with farmers' degree of concern about pollution from agricultural chemicals (Buttel *et al.*, 1981) or with farmers' attitudes toward regulation of animal feed additives and other agricultural chemicals (Hoiberg and Bultena, 1981; Gillespie and Buttel, 1989). Nor is educational attainment necessarily a good predictor of the adoption of conservation practices (Pampel and Van Es, 1977; Taylor and Miller, 1978; Chamala, Keith and Quinn, 1982, 1983). Nevertheless, because it is usually the second best sociodemographic predictor of environmental concern among the general public (Van Liere and Dunlap, 1980; Buttel, 1987; Jones and Dunlap, 1992), educational attainment will be included as an independent variable in the present chapter. The variations from one State to another in some of the findings of Buttel *et al.* (1981), Hoiberg and Bultena (1981), and Gillespie and Buttel (1989) suggest that State should also be included as an independent structural or institutional variable. In Australia, agricultural extension services are the primary responsibility of State government departments, with extension policies varying from one State to another. Most of the laws and regulations relating to farm or consumer health and safety are also State-based, although the Commonwealth of Australia has established a national system for evaluating the safety of agricultural and veterinary chemicals, on the basis of which such chemicals may be cleared for registration in the various States. Moreover, while farmer organizations are federated at a national level, they each draw their members mainly from a particular State. Consequently, farmers' attitudes may be influenced by factors specific to their particular State. Another structural or institutional variable whose influence needs to be considered in Australia is membership or non-membership of a landcare group. As was mentioned earlier, landcare groups are largely autonomous local groups of people, mainly land users in rural areas, who aim to deal with problems of land degradation and to develop more sustainable farming systems and land management practices. Black and Reeve (1993) have examined the factors which influence the likelihood of a farmer's being a member of a landcare group. In the present chapter, landcare group membership is treated as a dichotomous independent variable. Also worthy of consideration as an independent variable is whether or not a farmer has what is sometimes termed a 'whole farm plan' or 'property management plan' — that is, a physical property plan showing such information as the following: soil types, intended future layout of fences, laneways, cultivation and grazing, dams and watering points, shelter belts, wildlife corridors and refuges, and vegetative fire barriers. As the development of whole farm plans is advocated by agricultural extension services, landcare groups and conservationists, possession of such a plan is one indicator of a farmer's openness to ideas and information from such sources, as well as of his or her likely attitude to the very process of environmentally relevant planning.

This section has outlined the independent variables used in this study. The next section will outline how these variables, together with the dependent variables, were measured.

3. SURVEY DESIGN AND SAMPLE

In order to identify rural environmental issues on which farmers' attitudes would be sought, a review was made of topics which have surfaced in the press, as well as of literature on environmental policy instruments (e.g. Bohm and Russell, 1985; Common, 1990) and any relevant previous studies, such as those mentioned above. This information was used as a basis for preparing a pool of Likert-type attitude statements from which 100 were chosen to construct multi-item attitudinal scales. Although the total number of items in each scale varied, an attempt to was made to have an approximately equal number of positive and negative items, so as to guard against response-set. A further section was added to the survey questionnaire in order to obtain sociodemographic and farm situational information from respondents. The choice of questions was guided by a consideration of abovementioned variables that, on theoretical grounds, might help to explain attitudinal variation. Modifications to the questionnaire were made after each of three pilot surveys, the last of which used a sample of 383 farmers in the southern New England region of New South Wales. The final form of the questionnaire included 75 randomly ordered Likert-type items with the following response categories: Strongly agree, Mostly agree, Neutral or not sure, Mostly disagree, Strongly disagree. Table 1 contains a list of the *a priori* scales constructed from 57 of these items. The remaining 18 Likert-type items were designed to measure attitudes to particular rural environmental policy issues not covered by these scales.

Table 1 List of *a priori* scales used in the nation-wide survey.

Scale		Number of items	Cronbach's alpha
A	Perception of the seriousness of land degradation	4	0.52
B	Attitude toward the idea of society at large subsidising the costs associated with the prevention or amelioration of land degradation	4	0.42
C	Attitude toward the idea of zoning rural land for specific uses as a means of preventing land degradation	2	0.63
D	Attitude toward agricultural chemicals, especially synthetic fertilisers and pesticides	8	0.79
E	Attitude toward the need for training in the use of agricultural chemicals	2	0.29
F	Qualms about chemical residues in food	4	0.61
G	Conservation orientation	8	0.64
H	Awareness of the wider social and environmental contexts of farming	7	0.38
I	Attitude toward external advice or influence in farm decision-making	8	0.65
J	Profit orientation	4	0.30
K	Risk orientation	2	0.11
L	Political/economic orientation	4	0.36

For the nation-wide survey, systematic samples were drawn from farmer organization membership lists in all States of Australia except New South Wales, where a systematic sample from those persons listed in farming occupations in the Yellow Pages of the telephone directories was used. A response rate of 57 per cent was obtained, yielding a total of 2044 usable questionnaires. A one-page follow-up questionnaire seeking minimal biographical and farm details was mailed to non-responders. Of these, 39 per cent returned the follow-up questionnaire. Comparison of the responders to the full questionnaire with those who returned the one-page follow-up indicated the possibility of response sample bias in respect of only one variable, namely highest level of education, but the differences in this respect were not very substantial (for details, see Black and Reeve, 1993). Comparison of the distribution of farm sizes in the survey sample with the distributions published by the Australian Bureau of Statistics (1991) indicated that farms greater than or equal to 500 hectares were 1.54 times over-represented in the sample, while farms less than 500 hectares were correspondingly under-represented. Consequently, the survey data are representative mainly of the larger fully commercial properties that are responsible for the bulk of agricultural production in Australia. Further, because of the sampling frame used, it is likely that hobby farms are under-represented among the smaller farms in the sample.

4. RESULTS OF ANALYSIS

It can be seen from Table 1 that in the nation-wide survey the Cronbach's alpha for the *a priori* scales ranged from .11 to .79, indicating that not all the scales were equally reliable. Five of the scales, namely C, D, F, G, and I, had sufficiently high alphas to warrant further use in analysis. Various reasons could, in principle, be advanced for the low alphas on the remaining scales:

- particular items of a scale may have been misunderstood due to ambiguous or complicated wording;

- a given item may have embodied more than one issue, idea, concept or object, so that different respondents may be responding to a different aspect;

- the items constituting a particular scale may canvas too broad a range of issues rather than focusing on a unidimensional attitude; and

- there exist particular patterns of response to a set of items which while being inconsistent as an indication of a simple unidimensional attitude, may nevertheless reflect a logically feasible view about an issue.

Much of the effort in trialling the questionnaire was directed at avoiding problems involving the first two reasons described above. However, a number of items were introduced for the first time in the national survey, generally to replace items that had performed poorly in the last of the pilot surveys. The last two reasons above may be the cause of low alphas for several of the scales. For example, Scale H probably attempted to gauge attitudes to too diffuse a concept. Nonetheless, a number of the items in this scale have been found useful in other aspects of analysis. The circumstances surrounding the fourth reason listed above were investigated in detail using the results for Scale A from the last of the pilot surveys. Reeve and Black (1994) showed that in that particular case the relatively low alpha was due to patterns of responses which, while being inconsistent from the point of view of an attitudinal scale, could nevertheless be formed into groups that reflected plausible views about the seriousness of land degradation. These groups included 'past pessimists' — those who believe that land degradation is common on rural properties, that agricultural land is not generally in good shape, but that things were much worse in the past — and 'marginal optimists' who, while being concerned about land degradation, nevertheless firmly believe that marginal lands can be used for agriculture without damage.

Given that some of the *a priori* scales did not work as well as had been hoped, it was decided to use exploratory factor analysis with all 75 Likert-type items, in an endeavour to identify those groups of items which tend to elicit similar patterns of response from any one respondent. The procedure followed was that recommended by Tabachnick and Fidell (1989). Responses to the attitudinal items were screened for univariate and multivariate outliers and thirty-two respondents with z-scores of 4.0 or greater were removed. There were no multivariate outliers. After screening and the removal of cases with missing data, 1703 cases were available for factor analysis. Ten of the seventy-five attitude statements had squared multiple correlations of less than 0.2 and these were removed from the factor analysis as outlier variables. The Kaiser-Meyer-Olm measure of sampling adequacy was 0.91. Principal components analysis was used as a preliminary analysis to establish the range within which the number of factors might lie. Four criteria were used to establish this range:

- the number of eigenvalues greater than one,

- the scree plot of eigenvalues,

- the first appearance of doublet factors (factors based on only two statements), and

- the interpretability of factors.

These criteria established the range as being from four to eleven factors. Eight separate factor analyses (principal axis factoring and orthogonal rotation) with four through to eleven factors were examined for interpretability and the presence of doublet and singlet factors. The eight factor solution was selected as that which provided the most factors with no uninterpretable factors or factors with all loadings less than 0.4. To test whether oblique or orthogonal rotation of solutions was appropriate, principal axis factoring with eight factors was repeated with oblique rotation (direct quartimin). This solution produced three singlet factors, with the remaining factors appearing to have less clear interpretations. In the matrix of correlations between factors, all correlations were less than 0.3 with the exception of one correlation of 0.34 and one of 0.31. In the light of these findings, it was judged to be preferable to use orthogonal rather than oblique rotation. While an interpretative loading cutoff of 0.4 was used in preliminary analysis, loadings of 0.3 and above are given in Table 2 for each of the eight factors.

Table 2 Factor loadings with absolute values greater than, or equal to, 0.30 for principal factors extraction and varimax rotation. (The letters A to L denote to which of the scales listed in Table 1 statements belong. S.M.C. is squared multiple correlation. Cronbach's alpha is given both for items with loadings greater than 0.4 and for items with loadings greater than 0.3. The two alpha values are separated by a slash. Statements that load on more than one factor are footnoted and the loading on the other factor given in the footnote. For a complete table of loadings, see Reeve and Black, 1993).

Factor 1—Agricultural chemicals: benign or not. (Per cent of variance = 6.68; S.M.C. = 0.84; alpha = 0.84/0.84).

F	The dangers of chemical residues in agricultural produce have been greatly exaggerated.	0.64
D	There is too much talk about the harmful environmental effects of pesticides and not enough about their benefits.	0.63
D	Agricultural pesticides are a serious threat to public health.	−0.61
D	The dangers of environmental pollution from agricultural chemicals have been greatly exaggerated.	0.60
D	Agricultural chemicals create far more problems than they solve.	−0.59
F	Many of the fears expressed by consumers about chemical residues in food are quite unreasonable.	0.57
D	If you follow the manufacturer's directions, the agricultural chemicals now available will not harm your health.	0.55
D	Sustainable agriculture in my industry will probably always require extensive use of agricultural chemicals.	0.50
D	The pollution effects of fertilizers are quite unimportant compared to their benefits in increasing production[2].	0.45
	If less pesticides were used, the profitability of agricultural production would be reduced.	0.41
D	Agriculture today is too dependent on the use of agricultural chemicals.	−0.40
	There is no need for a tax on agricultural chemicals to finance research on safer alternatives.	0.34
E	Most farmers are skilled in the correct use of the agricultural chemicals for their industry.	0.32

Factor 2—Additional policy measures: needed or not. (Per cent of variance = 6.52; S.M.C. = 0.82; alpha = 0.74/0.80).

C	Poor or marginal country should be officially zoned to show what types of agriculture shall be permitted.	0.57
C	Zoning rural land, so that farmers are only allowed to use it according to its capability, will be necessary to prevent land degradation in some areas of Australia.	0.49
E	The best way to reduce the misuse of agricultural chemicals is for there to be courses on safety that are compulsory before you can use a chemical.	0.48
I	Farmers should have to provide Environmental Impact Statements before undertaking large developments like feedlots or major land clearing on their properties.	0.46
	Satellite photography and remote sensing should be more widely used to monitor whether land degradation is occurring on individual properties.	0.46
	Farmers should not receive the financial benefits of primary producer status (e.g. income averaging, sales tax exemptions on certain items) unless they are following recommended sustainable agricultural practices.	0.46

A	Some marginal types of country being used for agriculture in Australia will never be able to be farmed or grazed without badly damaging the land.	0.45
	Buffer zones around towns where there is a lot of crop spraying are essential to protect the health of townspeople.	0.43
	Investment in landcare is important to ensure future farm profitability[3].	0.41
	Too little is being done to educate farmers about damage done to the environment by some agricultural practices.	0.40
F	There needs to be more testing of all agricultural produce for harmful chemical residues[4].	0.38
G	All necessary soil conservation methods should be used, whatever the costs.	0.38
G	People who knowingly pollute the countryside are just as criminal as people who steal.	0.38
I	For the farmer who wants to farm sustainably, it is essential to have a whole farm plan prepared.	0.37
F	There are not enough penalties against those farmers whose produce contains chemical residues at levels known to be harmful.	0.36
B	Farmers should be levied to pay for environmental research and monitoring in their industry.	0.34
I	It is well worth seeking outside advice to help you decide how best to use your land.	0.32

Factor 3—Profit from farming: more important than the environment or not. (Per cent of variance = 5.00; S.M.C. = 0.76; alpha = 0.71/0.73).

J	The most important objective of farming is to maximise profit.	0.52
G	Farmers have a greater responsibility to produce food and fibre than they have to preserve the rural environment.	0.50
J	There is no point in adopting new practices unless they are more profitable.	0.50
	On the whole, the benefits of applying landcare practices on-farm are being greatly over-rated.	0.49
	Knowing about how agriculture affects the environment is not important to being a good farmer.	0.46
G	It is worth putting up with a small decrease in farm profits to protect the environment[5].	−0.44
H	On the whole, what a farmer does on his/her property has very little effect on other properties.	0.43
H	Compared to industries like mining and manufacturing, agriculture has very little impact on the environment.	0.34
H	Natural bush generally harbours weeds and vermin.	0.31

Factor 4—Outside expertise: irrelevant or not. (Per cent of variance = 4.02; S.M.C. = 0.68; alpha = 0.47/0.53)

I	The farmer is the best person to decide how land degradation problems on his/her farm should be tackled.	0.51
I	Whole farm plans prepared by government agencies are an unnecessary interference with farmers' rights to use their land as they see fit.	0.50
	When it comes to tackling land degradation, practical first-hand experience is much more use than scientific theory.	0.46
L	There should be less government regulation of business.	0.42

I	Farmers are the best persons to decide how much of their land should be cleared[6].	0.41
I	It would be a waste of effort for governments to subsidise the preparation of environmentally sound management plans for each farm[7].	0.33
I	Farming recommendations from government departments are generally worth following.	−0.30

Factor 5—The conservation movement: disapprove or not. (Per cent of variance = 2.74; S.M.C. = 0.62; alpha = 0.65/0.71).

G	Governments these days pay too much attention to the 'green movement'.	0.53
G	Some of the things conservationists are trying to protect are not worth worrying about[8].	0.50
G	On the whole, I approve the policies of the Australian Conservation Foundation (ACF)[9].	−0.43
	Environmentalists seldom take account of the economic implications of their policies.	0.38
G	The Australian Conservation Foundation (ACF) should confine its activities to the cities and not be concerned with rural issues[10].	0.37

Factor 6—Addressing land degradation: a community responsibility or not. (Per cent of variance = 2.18; S.M.C. = 0.66; alpha = 0.53/0.53).

B	Without financial assistance, there is little farmers can do to prevent land degradation occurring on their properties.	0.53
	Financial incentives should be made available to encourage farmers to use soil improving practices, e.g. rotation, stubble retention or deep ripping.	0.48
B	It is unfair to expect farmers to bear the cost of repairing land degradation on their properties.	0.44
B	It is unfair to expect people in towns and cities to contribute to the cost of preventing land degradation on rural lands.	−0.43

Factor 7—Agricultural land condition: an optimistic view or not. (Per cent of variance = 1.71; S.M.C. = 0.54; alpha = 0.61/0.61).

A	The land used for agriculture in Australia is in better condition than it has ever been[11].	0.43
A	Compared to what happened in the past, the amount of land degradation occurring in Australia now is relatively minor.	0.42

Factor 8—Sustainable agriculture: less productive or not. (Per cent of variance = 1.43; S.M.C. = 0.56; alpha = −/0.26).

Making agriculture more sustainable will also reduce levels of production.	0.42
We must accept slower economic growth in order to protect the environment.	0.34

With the exception of ten of the 59 items that have loadings above 0.3, each item loads on only one factor and each factor has two or more items loading on it. The ten items that each load on two factors generally combine the ideas expressed by the two factors. The interpretation and naming of the factors follows the procedure recommended by Rummel (1970) in specifying the end points of the dimension of attitude or opinion the factor is taken to represent. For example, the first factor is taken to represent an attitudinal dimension relating to agricultural chemicals, the end points of which are the views (a) that agricultural chemicals cause little harm beyond their targets, and (b) that agricultural chemicals do have serious unintended consequences. The statements after the colons in the factor labels specify the two endpoints, the first endpoint being that associated with a high score on the factor.

Factor 1 appears to encompass attitudes both to agricultural chemicals and to chemical residues. This suggests that the simultaneous holding of a favourable attitude to chemicals and an unfavourable one to chemical residues, or *vice versa*, is relatively uncommon. The statements that load on Factor 2 mostly relate to policy measures to reduce the environmental impacts of agriculture. The measures include: zoning of agricultural land; farmer education; cross-compliance; requirement to provide Environmental Impact Statements; monitoring land condition by remote-sensing; penalties for contamination with agricultural chemicals; and levies to fund environmental research and monitoring. The fact that none of these were formed into separate factors would suggest that attitudes to policy measures are fairly consistent across a wide range of instruments. The statements that load on Factor 3 span the issue of whether farming is strictly a business to which environmental concerns are largely irrelevant, or whether the environment, both on the farm and beyond its boundaries, is an important issue for the farmer. In contrast to Factor 2 which relates to policy instruments rather than their point of application, Factor 4 focuses on the latter aspect in the form of farmers' decision-making. The inclusion of 'There should be less government regulation of business' in this factor rather than in Factor 2 may be because responses were influenced by the form of regulation being unspecified and by respondents interpreting 'business' as their own farm business rather than the private sector in general. It should also be noted that whereas a high score on Factor 2 indicates approval of further government policy measures, a high score on Factor 4 indicates an assertion of farmers' competence to decide by themselves how to use their land. Factor 5 is probably the most unambiguously interpretable factor of the eight, clearly relating to attitudes to conservationists and the organizations that represent them. With the exception of one, the statements loading on this factor are a subset of the *a priori* attitudinal scale G (Conservation orientation). The remaining statements in this attitudinal scale load on Factors 2 and 3, which suggests that separate scales may be needed to distinguish attitude to conservationists from attitude to conservation. The statements that load on Factor 6 encompass the issue of whether farmers should be solely responsible for the rehabilitation and prevention of land degradation on their properties, or whether they can expect the assistance of society at large in these tasks. Factor 7 is a doublet factor that relates to whether land degradation in Australia can be viewed as a minor or insignificant problem (termed the optimistic view) or otherwise. Factor 8, also a doublet factor, relates to whether or not there is a trade-off between sustainability/environment and production/growth. The first three factors have a high degree of internal consistency, with squared multiple correlations between factor scores and scores on variables (i.e. attitude statements) of 0.76 or more. The fourth through to the eighth factor have decreasing internal consistency, the eighth factor having a squared multiple correlation of 0.56. A fairly similar gradation is found in the reliabilities (Cronbach's alpha) for each factor, as shown in Table 2. Multiple regression was used to ascertain the extent to which factor scores were correlated with, or could be predicted from, non-attitudinal variables. Some transformations of the non-attitudinal variables were necessary to provide a set of independent variables meeting the assumptions of multiple regression. Area of property, being positively skewed, was transformed to the logarithm of the area. The proportion of total property area in leasehold land was strongly skewed and platykurtic, and was therefore dichotomized at 0.5. Nominal variables of n categories were replaced with n-1 dichotomous variables. Variables were screened for univariate and multivariate outliers, the latter being identified using Cook's distance and standardized residuals. Dichotomous variables with less than five per cent of cases in one category were omitted from regressions as outlier variables. After screening and removal of cases with missing data, approximately 1450 cases were available for the regressions. Scatterplots and normal probability plots of residuals were used to check that assumptions of normality, linearity, homoscedasticity and independence of residuals had been met in the regressions accepted for interpretation. Stepwise linear regression was used as an exploratory technique to identify non-attitudinal variables making a significant contribution to the prediction of the dependent variable. These variables were then used in a standard multiple regression to simplify interpretation. Of the eight factors obtained in the factor analysis described above, Factors 1, 2, 3 and 4 had a multiple correlation coefficient (R) greater than 0.3. The results for these regressions are summarized in Table 3.

Table 3 Standard multiple regression of non-attitudinal variables on Factors 1 to 4.

Beta Values

	Factor 1- Agricultural chemicals: benign or not	Factor 2- Additional policy measures: needed or not	Factor 3- Profit from farming: more important than the environment or not	Factor 4- Outside expertise: irrelevant or not
Balance between grazing and cropping enterprises (all grazing = 1; all cropping = −1)	−0.212[a]			0.071[b]
Possesses a whole farm plan (2)		0.185[a]		
Member of a landcare group (2)			−0.183[a]	−0.093[a]
Highest level of formal education			−0.147[a]	−0.191[a]
Dairy farmer (2)				−0.152[a]
Number of years involvement in farming as an adult	0.131[a]		0.126[a]	0.142[a]
Extent of off-farm employment	−0.122[a]		−0.087[a]	
Female gender (2)	−0.121[a]			0.087[a]
Non-agriculturally derived income more than 50% of total income (2)				−0.073[b]
Logarithm of property area		−0.154[a]	−0.081[b]	
Age		0.130[a]		
Level of equity less than 50% (2)			0.064[b]	0.052[c]
Property located in Victoria (2)		−0.117[a]	0.063[c]	
Cereal grower (2)		−0.111[a]		
Fodder grower (2)	−0.080[b]			
Grain legume grower (2)	0.067[c]			
Farm business structured as a family partnership (2)	0.057[c]	−0.053[c]	−0.058[c]	
R	0.36[a]	0.32[a]	0.37[a]	0.34[a]
R Squared	0.13	0.10	0.13	0.12
Adjusted R Squared	0.13	0.10	0.13	0.12
N	1453	1454	1456	1455

Note 1: Binary variables are denoted by (2) after the description. The category described is that taking the value of one in the regression.

Note 2: Significance levels — a: $p<0.001$, b: $p<0.01$, c: $p<0.05$

It can be seen from Table 3 that the type of agricultural production in which one is engaged tends to influence one's attitude to some issues. Those engaged in cropping enterprises, other than production of fodder crops, tend to perceive fewer environmental and health problems associated with the use of agricultural chemicals (Factor 1) than do graziers. Whilst this difference may be due to croppers' greater familiarity with, and proficiency in, safe chemical use, it may also be due simply to the fact that they depend far more on agricultural chemicals, whether there are hazards associated with these or not. Cereal growers in particular tend to be wary of the possibility of additional policy measures aimed at protecting the environment (Factor 2). Dairy farmers are more likely than others to affirm the value of expertise from outside the farm (Factor 4). Higher levels of education and membership of a landcare group each tend to be associated with more favourable attitudes toward expertise from outside the farm (Factor 4) and also with a lower likelihood of one's affirming that profit from farming is more important than the environment (Factor 3). The analysis presented in Black and Reeve (1993) suggests that, at least in the early stages of the landcare movement, such attitudes may be more a cause than a consequence of landcare group membership. A longitudinal study is, however, needed to ascertain to what extent landcare group membership results in changes of attitude and behaviour with respect to the environment. Another important predictor of attitudes is the number of years in adulthood a person has been involved in farming. This variable tends to be positively associated with a favourable attitude toward agricultural chemicals (Factor 1), with an opinion that profit from farming is more important than the environment (Factor 3), and with a reluctance to draw on expertise from outside the farm (Factor 4). Years of farm experience as an adult is not simply a reflection of age, as the latter is not significantly associated with any of these attitudes or opinions when one controls statistically for other non-attitudinal variables (including years of farm experience as an adult). With similar statistical controls, age tends, however, to be positively associated with a willingness to accept additional policy measures in order to achieve environmentally beneficial objectives (Factor 2), whereas years of farm experience as an adult is not. Of the items with significant loadings on Factor 2, the following are the two on which age differences are most evident: 'Some marginal types of country being used for agriculture in Australia at the moment will never be able to be farmed or grazed without badly damaging the land' and 'Poor or marginal country should be officially zoned to show what types of agriculture shall be permitted'. Amongst farmers aged 60 years or more, the proportions agreeing with these propositions are 68.5 per cent and 60.8 per cent respectively, whereas the corresponding proportions among farmers aged less than 30 years are 50.5 per cent and 45.6 per cent respectively. These differences may reflect a greater awareness among older farmers than among younger farmers of previous government intervention on land use, especially in closer settlement schemes. The greater the extent of a farmer's off-farm employment, the greater the likelihood that he or she will have qualms about agricultural chemicals (Factor 1), and the lower the likelihood that he or she will affirm that profit from farming is more important than the environment (Factor 3). These results could, in principle, be due to the fact that off-farm employment tends to increase one's contact with non-farmers, whose attitudes on such matters may differ from those held by persons whose sole occupation is farming; also to the fact that off-farm income may enable one to maintain a desired standard of living without using environmentally damaging agricultural practices. However, non-agriculturally derived income as a proportion of total net income adds little to the explanation of farmers' scores on Factors 1 and 3 after one has taken account of the other variables listed in Table 3. By contrast, it contributes significantly to the explanation of favourable attitudes toward expertise from outside the farm (Factor 4). Women, who make up 6 per cent of the sample, tend to be more apprehensive than men both about agricultural chemicals (Factor 1) and about the value of expertise from outside the farm (Factor 4). Both these attitudes could be interpreted as evidence of conservatism among female landholders with respect to farming practices. Also, the greater apprehension about chemicals is consistent with the tendency for concerns about health and safety issues to be raised more frequently among rural women's groups than by farmer organizations. Those with larger landholdings — some of whom are in temperate zones but many of whom are in parts of the country which are most ecologically fragile — are more likely than others to oppose additional policy measures aimed at protecting the environment (Factor 2). On the other hand, they are less likely than others to hold that profit from farming is more important than the environment (Factor 3). Other things being equal, large properties are more likely to be profitable than are smaller properties, and for this reason persons with large farms may be less inclined to adopt an exploitative attitude toward the environment. Nevertheless, they tend to oppose any encroachment on their autonomy. Respondents from the State of Victoria tend to have a less favourable attitude than those in other States to additional policy measures aimed at protecting the environment (Factor 2). In this respect they are similar to those with large landholdings. But they differ from the latter in that they are more likely than others to hold that profit from farming is more important than the environment (Factor 3). Further research is needed to establish why such opinions are more widely held among farmers in Victoria than in other States. Explanations to be investigated could include explicit stances taken by State-based farmer organizations, particular policies pursued by the various State governments and their political opponents, and differential media coverage of rural environmental issues. Those with less than 50 per cent equity in their landholding are more likely than others to regard profit from farming as more important than the environment (Factor 3); they are less likely than others to look favourably on expertise from outside the farm (Factor 4). These attitudes are understandable, given the imperatives of such persons to service their debts, many of which were contracted on the basis of advice that did not anticipate unfavourable movements in commodity prices and interest rates. Finally, persons whose farm business is structured as a family partnership are more likely than others to have a favourable attitude toward agricultural chemicals (Factor 1) and to oppose additional policy measures aimed at protecting the environment (Factor 2). Despite this, they are less likely than others to regard profit from farming as being more important than the environment (Factor 3). Structuring the farm business as a family partnership is generally advantageous for taxation purposes if neither partner has substantial off-farm employment or off-farm income. Consequently, adopting such a structure is often regarded as a evidence of a greater level of financial skill than is sole proprietorship. Just over two-thirds of the farms in our sample were family partnerships, 16 per cent were sole proprietorships, and 13 per cent were operated as public companies. Although 'sole proprietorship' and 'public company' were also entered into the initial regression equations as binary variables, neither of these variables had a significant association with Factors 1 to 4 when one controlled statistically for the other variables in the final regression equations. Thus, contrary to what has sometimes been suggested, our data show no evidence that operators of farms structured as public companies are less concerned than other farmers about protecting the environment.

5. CONCLUSIONS

This chapter has focused on some of the main findings from a large-scale study of Australian farmers' attitudes toward environmental issues. From a pool of 75 Likert-type items it was possible to identify eight orthogonal attitudinal factors or dimensions, each with its own set of indicators. These factors or dimensions measured farmers' perceptions of the following:

1. Agricultural chemicals: benign or not.
2. Additional policy measures: needed or not.
3. Profit from farming: more important than the environment or not.
4. Outside expertise: irrelevant or not.
5. The conservation movement: disapprove or not.
6. Addressing land degradation: a community responsibility or not.
7. Agricultural land condition: an optimistic view or not.
8. Sustainable agriculture: less productive or not.

On the basis of prior theorizing and research, it was hypothesized that various non-attitudinal variables would account for some of the attitudinal variations. Multiple regression models showed that non-attitudinal variables could explain more than 10 per cent of the variance on each of the first four factors, but that the particular combination of non-attitudinal variables was different for different factors. Listing the non-attitudinal variables in each case in order of importance, it was found that:

- Attitudes toward agricultural chemicals are more favourable among those with a predominance of cropping enterprises, those with more farming experience, those with no off-farm employment, males, non-fodder-growers, grain legume producers, and those whose farm business is structured as a family partnership.

- Attitudes to additional policy measures designed to protect the environment are more favourable among those with a whole farm plan, those with smaller farms, older farmers, those outside the State of Victoria, those who are not cereal growers, and those whose farm business is not structured as a family partnership.

- Attitudes to the balance between farming for profit and farming to maintain environmental quality are more in favour of profit among those who are not members of landcare groups, those who have lower levels of formal education, those with more farming experience, those with no off-farm employment, those with smaller farms, those with low equity in their farms, those in Victoria, and those whose farm business is not structured as a family partnership.

- Attitudes toward outside expertise relevant to farm decision making are more favourable among those with higher levels of formal education, dairy farmers, those with less farming experience, landcare group members, males, those for whom non-agriculturally derived income constitutes more than half of their total net income, those with a predominance of cropping enterprises, and those with higher levels of equity in their farm.

Whilst most of the above results are consistent with what had been hypothesized on theoretical grounds, some hypothesized associations were not evident when one controlled statistically for other variables. Thus, although the initial regression models included a variable measuring the extent to which a farm operator was using leasehold land rather than freehold land, this variable had no significant association with Factors 1 to 4 when one controlled for the other variables in the final regression equations. The same can be said of a variable measuring whether the farm business was operated as a public company, and of another variable measuring whether any part of the farm property had previously been owned by a parent of either the farm operator or the farm operator's spouse. Furthermore, as already noted, other non-attitudinal variables had significant associations with some of the attitudinal factors but not with others. Possible reasons for such differences have been discussed in the chapter. Some of the results were actually the opposite of what has previously been hypothesized. Thus, contrary to what some writers have suggested, operators of large properties are *less* likely than others to hold that profit from farming is more important than the environment. Nevertheless, they are more likely than others to oppose additional policy measures aimed at protecting the environment. In this combination of attitudes, large landholders are similar to persons operating their farms as family partnerships. Clearly, constellations of attitudes and opinions are more complex than has sometimes been assumed. The fact that non-attitudinal variables explain only a relatively small proportion of the variance in the attitudes measured by Factors 1 to 4 is broadly consistent with the findings in previous studies of environmental concern among the public at large (Van Liere and Dunlap, 1990; Jones and Dunlap, 1993).

REFERENCES

Arcury, T.A. and E.H. Christianson. (1990). 'Environmental Worldview in Response to Environmental Problems: Kentucky 1984 and 1988 Compared'. *Environment and Behavior* 22:378-407.

Australian Bureau of Statistics. (1991). *Agricultural Industries: Structure of Operating Units, Australia, 31 March 1990*. Canberra: Australian Bureau of Statistics Cat. No. 7102.0.

Barlett, P.F. (1986). 'Part-time Faming: Saving the Farm or Saving the Life-Style?'. *Rural Sociology* 51:289-313.

Black, A.W. and I. Reeve. (1993). 'Participation in Landcare Groups: the Relative Importance of Attitudinal and Situational Factors'. *Journal of Environmental Management* 39:51-71.

Blocker, T.J. and D.L. Eckberg. (1989). 'Environmental Issues as Women's Issues: General Concerns and Local Hazards'. *Social Science Quarterly* 70:586-593.

Bohm, P. and C.S. Russell. (1985). 'Comparative Analysis of Alternative Policy Instruments'. In: A.V. Kneese and J.L. Sweeney (eds), *Handbook of Natural Resource and Energy Economics*. Amsterdam: Elsevier, p. 395-460.

Brody, C.J. (1984). 'Differences by Sex in Support for Nuclear Power'. *Social Forces* 63:209-228.

Bultena, G.L. and E.O. Hoiberg. (1983). 'Factors Affecting Farmers. Adoption of Conservation Tillage'. *Journal of Soil and Water Conservation* 38:281-284.

Buttel, F.H. (1987). 'New Directions in Environmental Sociology'. *Annual Review of Sociology* 13:465-488.

Buttel, F.H. and W. Flinn. (1974). 'The Structure of Support for the Environmental Movement, 1968-1970'. *Rural Sociology* 39:56-59.

Buttel, F.H. (1987). 'New Directions in Environmental Sociology'. *Annual Review of Sociology* 13:465-488.

Buttel, F.H., G.W. Gillespie Jr., O.W. Larson and C.K. Harris. (1981). 'The Social Bases of Agrarian Environmentalism: A Comparative Analysis of New York and Michigan Farm Operators'. *Rural Sociology* 46:391-410.

Buttel, F.H. and G.W. Gillespie Jr. (1984). 'The Sexual Division of Farm Household Labor: An Exploratory Study of On-Farm and Off-Farm Labor Allocation Among Farm Men and Women'. *Rural Sociology* 49:183-209.

Cameron, J.I. and J. Elix. (1991). *Recovering Ground. A Case Study Approach to Ecologically Sustainable Rural Land Management*. Melbourne: Australian Conservation Foundation.

Campbell, A. (1991). *Planning for Sustainable Farming*. Port Melbourne: Lothian Publishing Company.

Chamala, S., K.J. Keith and P. Quinn. (1983). 'Australian Farmers' Attitudes towards, Information Exposure to, and Use of Commercial and Soil Conservation Practices'. *Tillage Systems and Social Science* 3(1):1-3.

Common, M.S. (1990). 'Policy Instrument Choice'. In: *Moving Toward Global Sustainability: Policies and Implications for Australia*. Canberra: Australian

National University, p. 87-116.
Commonwealth of Australia. (1990). *Ecologically Sustainable Development*. A Commonwealth Discussion Paper. Canberra: Australian Government Publishing Service.
Coughenour, C.M and L. Swanson. (1983). 'Work Statuses and Occupations of Men and Women in Farm Families and the Structure of Farms'. *Rural Sociology* 48:24-43.
Earle, T.R., C.W. Rose and A.A. Brownlea. (1979). 'Socioeconomic Predictors of Intention Towards Soil Conservation and their Implication in Environmental Management'. *Journal of Environmental Management* 9:225-236.
Ervin, C.A. and D.E. Ervin. (1982). 'Factors Affecting the Use of Soil Conservation Practices: Hypotheses, Evidence, and Policy Implications'. *Land Economics* 58:277-292.
Ervin, D. (1986). 'Constraints to Practicing Soil Conservation: Land Tenure Relationships'. In: S.B. Lovejoy and T.L. Napier (eds), *Conserving Soil: Insights from Socioeconomic Research*. Ankeny, Iowa: Soil Conservation Society of America, p. 95-120.
Fassinger, P.A. and H.K. Schwarzweller. (1984). 'The Work of Farm Women: A Midwestern Study'. In: H.K. Schwarzweller (ed.), *Research in Rural Sociology and Development*, Vol. 1. Greenwich, Conn.: Jai Press, p. 37-60.
Freudenburg, W.R. (1991). 'Rural-Urban Differences in Environmental Concern: a Closer Look'. *Sociological Inquiry* 61:167-198.
George, D.L. and P.L. Southwell. (1986). 'Opinion on the Diablo Canyon Nuclear Power Plant: The Effects of Situation and Socialization'. *Social Science Quarterly* 67:722-735.
Gillespie, G.W. Jr. (1987). 'Antibiotic Animal Feed Additives and Public Policy: Farm Operators' Beliefs about the Importance of these Additives and their Attitudes toward Government Regulation of Agricultural Chemicals and Pharmaceuticals'. Unpublished PhD dissertation, Cornell University.
Gillespie, G.W. Jr. and F.H. Buttel. (1989). 'Understanding Farm Operators' Opposition to Government Regulation of Agricultural Chemicals and Pharmaceuticals: The Role of Social Class, Objective Interests, and Ideology'. *American Journal of Alternative Agriculture* 4(1):12-21.
Glenn, N.D. and J.P. Alston. (1977). 'Rural-Urban Differences in Reported Attitudes and Behaviors'. *Southwestern Social Science Quarterly* 47:381-400.
Hamilton, L.C. (1985a). 'Concerns About Toxic Wastes: Three Demographic Predictors'. *Sociological Perspectives* 28:463-486.
Hamilton, L.C. (1985b). 'Who Cares About Water Pollution? Opinions in a Small-Town Crisis'. *Sociological Inquiry* 55:170-181.
Hawkes, G.R., M. Pilisuk, M.C. Stiles and C. Acredolo. (1984). 'Assessing Risk: A Public Analysis of the Medfly Eradication Program'. *Public Opinion Quarterly* 48:443-451.
Hoiberg, E.O. and G.L. Bultena. (1981). 'Farm Operator Attitudes Toward Governmental Involvement in Agriculture'. *Rural Sociology* 46:38-390.
Honnold, J.A. (1981). 'Predictors of Public Environmental Concern in the

1970s'. In: D. Mann (ed.), *Environmental Policy Formation*. Vol. 1. Lexington: Lexington Books, p. 63-75.

Honnold, J.A. (1984). 'Age and Environmental Concern: Some Specification of Effects'. *Journal of Environmental Education* 16:4-9.

Jacks, G.V. and R.O. Whyte. (1939). *The Rape of the Earth: A World Survey of Soil Erosion*. London: Faber.

Jones, R.E. and R.E. Dunlap. (1992). 'The Social Bases of Environmental Concern: Have They Changed over Time?' *Rural Sociology* 57:28-47.

Lee, L.K. (1980). 'The Impact of Landownership Factors on Soil Conservation'. *American Journal of Agricultural Economics* 62:1070-1076.

Lee, L.K. and W.H. Stewart. (1983). 'Landownership and the Adoption of Minimum Tillage'. *American Journal of Agricultural Economics* 65:256-264.

Lowe, G.D. and T.K. Pinhey. (1982). 'Rural-Urban Differences in Support for Environmental Protection'. *Rural Sociology* 47:114-128.

Lynne, G.D., J.S. Shonkwiler and L.R. Rola. (1988). 'Attitudes and Farmer Conservation Behavior'. *American Journal of Agricultural Economics* 70:12-19.

McStay, J.R. and R.E. Dunlap. (1983). 'Male-Female Differences in Concern for Environmental Quality'. *International Journal of Women's Studies* 6:291-301.

Mohai, P. (1992). 'Men, Women and the Environment: An Examination of the Gender Gap in Environmental Concern and Activism'. *Society and Natural Resources* 5:1-19.

Mohai, P. and B.W. Twight. (1987). 'Age and Environmentalism: an Elaboration of the Buttel Model Using National Survey Evidence'. *Social Science Quarterly* 68:798-815.

Mohai, P. and B.W. Twight. (1986). 'Rural-Urban Differences in Environmentalism Revisited: Nature Manipulative Vs Nature Exploitative Orientations'. Paper presented at the Annual Meeting of the Rural Sociological Society.

Norris, P.E. and S.S. Batie. (1987). 'Virginia Farmers' Soil Conservation Decisions: An Application of Tobit Analysis'. *Southern Journal of Agricultural Economics* 19:79-90.

Pampel, F. Jr. and J.C. van Es. (1977). 'Environmental Quality and Issues of Adoption Research'. *Rural Sociology* 42:57-71.

Passino, E.M and J.W. Lounsbury. (1976). 'Sex Differences in Opposition to and Support for Construction of a Proposed Nuclear Plant'. In: L. M. Ward et al. (eds.), *The Behavioral Basis of Design*, Book 1. Stroudsburg, Pa.: Dowden, Hutchinson and Ross, p. 1-5.

Pfeffer, M.J. (1992). 'Labor and Production Barriers to the Reduction of Agricultural Chemical Inputs'. *Rural Sociology* 57:347-362.

Powell, J.M. (1976). *Environmental Management in Australia 1788-1914*. Melbourne: Oxford University Press.

Reeve, I.J. and A.W. Black. (1993). *Australian Farmers' Attitudes to Rural Environmental Issues*. Armidale, N.S.W.: The Rural Development Centre,

University of New England.

Reeve, I.J. and A.W. Black. (1994). 'Understanding Farmers' Attitudes to Land Degradation: Some Methodological Considerations'. *Land Degradation and Rehabilitation* 5(4).

Roberts, B. (1992). *LandCare Manual*. New South Wales Kensington: University Press.

Rosenfeld, R.A. (1985). *Farm Women: Work, Farm and Family in the United States*. Chapel Hill, N.C.: University of North Carolina Press.

Rummel, R.J. (1970). *Applied Factor Analysis*. Evanson, Ill: Northwestern University Press.

Simpson, E.H., J. Wilson and K. Young. (1988). 'The Sexual Division of Farm Household Labor: A Replication and Extension'. *Rural Sociology* 53:145-165.

Solomon, L.S., D. Tomaskovic-Devey and B.J. Risman. (1989). 'The Gender Gap and Nuclear Power: Attitudes in a Politicized Environment'. *Sex Roles* 21:401-414.

Steger, M.A.E. and S.L. Witt. (1989). 'Gender Differences in Environmental Orientations: A Comparison of Publics and Activists in Canada and the U.S.'. *Western Political Quarterly* 42:627-649.

Stout-Wiegand, N. and R.B. Trent. (1983). 'Sex Differences in Attitudes to New Energy Resource Developments'. *Rural Sociology* 48:627-646.

Tabachnick, B.G., and L.S. Fidell. (1989). *Using Multivariate Statistics*. New York: Harper and Row.

Taylor, D.L. and W.L. Miller. (1978). 'The Adoption Process and Environmental Innovations: A Case Study of a Government Project'. *Rural Sociology* 43:634-648.

Tremblay, K.R. Jr. and R.E. Dunlap. (1978). 'Rural-Urban Residence and Concern with Environmental Quality: A Replication and Extension'. *Rural Sociology* 43:474-491.

Van Liere, K. D. and R.E. Dunlap. (1980). 'The Social Bases of Environmental Concern: a Review of Hypotheses, Explanations and Empirical Evidence'. *Public Opinion Quarterly* 44:181-197.

Wimberley, R.C. (1983). 'The Emergence of Part-Time Farming as a Social Form of Agriculture'. In: I.H. Simpson and R.L.Simpson (eds), *Research in Sociology of Work: Peripheral Workers,* Vol. 2. Greenwich, Conn.: Jai Press, p. 325-356.

World Commission on Environment and Development. (1987). *Our Common Future*. London: Oxford University Press.

NOTES

1. This research was supported by the Rural Industries Research and Development Corporation. The assistance of the farmer organizations that enabled us to survey their members is gratefully acknowledged.

2. 0.32 on Factor 3.

3. −0.32 on Factor 3.

4. −0.32 on Factor 1.

5. 0.35 on Factor 2.

6. 0.32 on Factor 3.

7. −0.31 on Factor 3.

8. 0.31 on Factor 3.

9. 0.37 on Factor 2.

10. 0.31 on Factor 3.

11. 0.30 on Factor 3.

11. Methodological Problems of Measurement of Ecological Attitudes and Comparison of Survey Data

Vladimir Rukavishnikov

Abstract

Albert Einstein wrote that 'the formulation of a problem is often more essential than its solution'. This chapter does not concentrate on what scholars actually know about public concerns and attitudes towards the environment, but rather on advantages and limitations of contemporary methodological approaches and surveys as sources of scientific and policy relevant information on mass ecological attitudes. The analysis is based on an overview of what has changed in the scientific knowledge about public opinion regarding environmental issues since the 1960's. A set of basic research questions, explored in the publications of recent years, as well as some answers sociologists have got on them, are formulated and examined in the framework of general problems of polling, comparison and generalization of empirical findings.

1. INTRODUCTION

In the mid-1960's environmental issues became one of the most debatable themes in contemporary world media. The public concern about environmental problems now appears on the list of most urgent issues in virtually every nation. However, this chapter is about the methodological aspects of the study of public opinion regarding environmental issues (popular concerns, attitudes, willingness to act, etc.), rather than about public views on the environment itself. I will not concentrate on what we actually know about public concerns and attitudes in this field, but on what has really changed in the scientific knowledge about the public opinion towards environmental issues since the 1960's, due to advantages of contemporary methodological approaches and some 'natural' limitations of public opinion polls as sources of information. This is the focal point of my discussion. There is an impressive amount of available texts, describing the public opinion in this field in a particular country as well as in cross-national perspectives. But I do not concern myself with an overview of these texts here. The references cited below are only those that are thought to deal especially with the subject matter of this chapter or give additional information about the topics introduced in the given section of the chapter.

There is a very large number of articles and books that examine public opinion about environmental issues that social scientists have been publishing about over

the years. Most authors of these articles and books consider the following research questions:
(1) Who is most concerned?
(2) What prompts environmental concerns?
(3) How willing are people to act?
(4) What do people think should be done?

The value and importance of these questions, which were already addressed in the analytic review of surveys carried out in the 1970's (Lipsey, 1977), are obvious from both theoretical and practical points of view and have not changed since. The principal shift I want to emphasize is the following: by the beginning of the 1990's, environmental problems became policy and political problems in most nations (the World Summit in Rio de Janeiro in 1992 displayed this), and currently governmental contracts are influencing research priorities in at least the same degree as the personal scientific interests of the investigators. These four questions should be regarded as basic research questions in almost all kinds of environmental surveys all over the world.

The *descriptive* character of the answers to most of the above mentioned research questions is obvious. Naturally, this character of research orientation first of all depends on the scientific tasks and practical problems which scholars, politicians and pollers deal with in every particular case. The fundamental methodological aspect of descriptive studies pays much more attention to *what* is happening than to *why* this is happening. However, it is obvious that today a pure description of facts no longer suffices, and therefore we must go beyond these facts.

Certainly, everybody who studied methodology can argue that description and explanation are two sides of the same coin, and what one scholar calls description may be called explanation by another. A characteristic singled out to describe a phenomenon is almost by definition one that the analyst considers to be of importance in attempting to explain the phenomenon. If this characteristic is not of importance, there is little reason for mentioning it: indeed, what the analyst 'sees' is itself usually derived from what he considers to be of explanatory importance.

2. ADVANTAGES AND LIMITATIONS OF SURVEYS' INFORMATION ABOUT ECOLOGICAL ATTITUDES

In almost all developed countries, pollers regularly provide decision makers, the general public and scholars with up-to-date facts and figures regarding the state of public opinion towards environmental issues. The important point is whether public opinion polls aiming at getting empirical answers to the four above stated questions, really help to find these answers. There are several reasons to doubt the efficiency of public opinion polling regarding these tasks.

Before proceeding to further analysis, I would also like to note the following. There is also doubt about whether the individual's responses in polls can be taken seriously as a genuine environmental attitude measurement. This view stands on a well-known and generally accepted definition of an attitude as a broad evaluative orientation of the respondent towards a particular class of objects. Attitudes deal with unverifiable matters involving preferences, while opinions, as statements indicating a person's subjective view on the issue of interest, deal with matters which are factual. Public opinion refers to the shared opinions of large groups of people. Though it is dangerous to generalize, one can still raise the question what has actually been studied in public opinion polling.

Considering the first and second basic research question, we should very well realize the difference between the collective public opinion and the opinion of the so-called 'issue public'. An issue public might be defined as those individuals in the public that possess quite definite attitudes towards particular issues. In almost all studies that are based on polls' data and focused on the above mentioned research questions, scholars deal almost exclusively with the environmental issue public, which, according to me, are those respondents that hold attitudes concerning the state of nature and related topics. These individuals need not be members of formal associations which advocate the importance of their particular core value, such as 'the Green parties', 'Greenpeace', etc. It should be recognized that because an opinion is offered, for example, about a government's environmental spending, this does not mean that an ecological attitude exists. It may simply be a response to a certain stimulus, in this case a poll. Interpreting the proportions of the collective public opinion on certain environmental issues, we should not forget that respondents to a particular question do not necessarily all have to be members of the environmental issue public. Polls record the total distribution of responses to questions, and depending upon the validity of the instrument, more or less touch upon the issue public. Here we come to the problem of the validity of surveys' information and instruments, i.e. its ability to measure the public's environmental attitudes. Assuming that polls are based on an accurate representative sample, I am talking here about validity in the general sense of the word and do not define it in terms of statistical criteria.

It should be emphasized that the researchers' conclusion about people's attitudes and opinions towards environmental issues is evidence taken from the individual's self-report of his own attitude and evaluation, and in drawing this conclusion, the researcher has to decide whether the respondent is truly aware of his own attitude and reports it accurately. The respondent's answers to the questionnaire items might be determined not only by the facts but also by his values and the impact of the media, which might determine the importance of the facts for him. With respect to environmental concerns as well as health matters (there is a strong link between these two topics), since there is nearly universal agreement on the desirability of protection of natural environment or decreasing of environmental degradation, the importance of values as well as facts in determining responses and actual behaviour is often overlooked. The fact that a non-polluted environment is a prerequisite for the population's health and further development of human activity, may by some respondents be taken as a basis for supporting projects that are oriented to conservation of the status-quo in the respondent's area of residence, and by others as a basis for supporting projects that are oriented to reconstruction of the environment or halting the population growth in other parts of the world.

In other words, searching for the answers to the questions 'who are most concerned' and 'what prompts environmental concerns', one should keep in mind the well-known fact that responses may in some cases reflect an individual's attitude, but they may also merely indicate a person's awareness of the common pattern of response. This can lead to misinterpretation of data at the aggregate level regarding the actual state of the collective public opinion and the opinion of the issue public. There is a huge gap between people's declarations of concern and their willingness to act on the one hand, and their behaviour on the other, which is registered in almost every study on environmental attitudes and behavior. The analysis of the diversity of responses to the question regarding the individual's willingness to donate money to prevent further pollution, published by Dutch scholars (Ester et al., 1993), might strengthen the position of those doubts in polls' data as a basis for the answers to the third and fourth basic research question. These authors state that conclusions and policy decisions based on the generalization of the findings describing the environmental attitudes spread among public, and the collection of people's views on what should be done in the area of environmental policy, might too often be misleading. It is true. Compared with the experts or even the issue public, the general public is not well-informed about and competent enough in policy-making.

I have to agree with the above presented critical judgements showing the limitations of information from surveys, but, in my view, starting from this basis of arguments, one cannot derive the general conclusion that information from polls is useless and invalid. We should not underestimate or overestimate the validity of public opinion surveys, and keep in mind that there is no other way to create a database of public views for both scientific inferences and political decision making.

Environmental surveys and policy-making

Public opinion polling, in the context of policy-making, should not be regarded as a purely scientific instrument, but as a policy making instrument. At best its purpose is to describe the population's (or population segment's) opinion and attitudes towards policy relevant issues, i.e. to produce additional arguments for policy-makers according to their demands. But are the policy-makers genuinely interested in what kinds of facts and factors have impacted the state and shifts in the opinions of voters?

In his recent study, the Canadian scholar Glen C. Filson repeated the conclusion drawn by Van Liere and Riley Dunlap in the 1980's that 'policy makers still know too little about the demographic and social factors influencing people, farmers in particular, to become concerned about environmental degradation' (Filson, 1993:166). He came to the same conclusion as his American colleagues, who, analyzing this issue in correlation with the factors determining New York farmers' preferences for low input production practices, have pointed out that 'little attention has been devoted to understanding the degree to which rank-and-file [North]American farmers now prefer or can, at some future point, be motivated to prefer lower input, more sustainable agricultural production systems'.

There are several reasons for such behaviour of policy-makers. The quality of information about environmental attitudes is one of them. Of course, the relationship between scientific research and policy towards the environment cannot simply be reduced to a technical matter. I think, however, merely describing the proportions of respondents who revealed different degrees of environmental concerns and/or willingness to offer part of their income to prevent further pollution, is not enough to explain peculiarities in the behaviour and opinion of different categories of people and policy decisions based on such information.

In the descriptive studies, the explanatory theory is usually presented in a very unexplicit form. For instance, when Glen Filson, who has been cited above, was trying to find a satisfactory interpretation frame for differences in environmental attitudes of Ontario farmers, he examined people's attitudes as a function of their age, education, sex, social class, income, occupation, farm size and the commodity of the produce (Filson, 1993). All these variables were used as explanatory ones. This is an example of one of the most frequently used methods of attitude research, when the conclusion confirming the theory is inferred through the analysis of the degree of statistical association between measures of environmental attitudes (concerns about protecting the environment and so on) and demographic, economic and social characteristics of respondents. Going forward along this path of searching for the basic research questions mentioned above, the researcher could get fairly enough information about groups of people that are more or less receptive to environmental issues and less to what prompts their environmental concerns and how the situation should be improved, i.e. information that is necessary for policy-making. And maybe in the origin of the opinion surveys' data, it is possible to find some causes making the process of implementation of results of environmental surveys into practical policy more complex and complicated than scholars desire to see it.

3. FOUR BASIC RESEARCH QUESTIONS FOR COMPARATIVE STUDIES

The cross-national studies, carried at the beginning of 1990s, were mainly focused on the following questions, which are a logical continuation of the first four basic research questions:

(5) What explains variations in environmental attitudes?
(6) What are the peculiarities of the relationship between environmental concerns, the willingness to act and the actual behaviour of populations of different countries?
(7) Who, according to the vast majority of populations of wealthy and poor

countries, are responsible for the deterioration of the national environment at local and global levels?

(8) What is the common conceptual pattern for the explanation of observed similarities and peculiarities of public opinion regarding environmental issues in different countries?

The answers to these questions cannot be obtained without knowledge of the responses to the first group of descriptive issues.

Although fundamental research questions and conceptual problems in comparative studies basically do not differ from those used for research in a single nation case, the multicultural setting of survey research greatly increased the magnitude of difficulties. Experience of recent years showed that such a widely accepted methodological demand as the use of identical questionnaires should be carefully discussed in each particular case. Because sometimes even the simple translation of questions from one language to another created misunderstanding of the issue by the respondent, due to socio-cultural differences between nations. This misunderstanding led to artifacts that could not easily be recognized by comparative analysts. Let me give an example to illustrate this thesis.

The Gallup 1992 world-wide environmental attitudes survey presented a fairly large variety of views on which nation is most concerned about environmental problems in their community, which actors people considered to be responsible for the deterioration of local and global natural environment, and what should be done to improve the situation. Despite the apparent similarities between the collective opinion of citizens of poor and wealthy countries all over the world, there were a number of differences in the perception of environmental issues by respondents from countries with similar levels of economic development as well as by representatives of similar cultures (Dunlap, et al., 1992). They attempted to interpret these findings in the framework of the modernization/postmodernization theory, emphasizing the role of differences in GNP per capita between rich and poor countries (see also chapter eight in this volume). However, analyzing the results of the Gallup world survey, the French sociologist Denis Duclos stressed the idea that because people from different nations tend to react to the same question according to various cultural patterns, it is also possible to link the observed diversity of the results of national surveys to the methodological aspects of this survey. And because the wordings used in the questionnaire were linked to the basic theoretical concepts, the instruments used in this global study have to be re-examined carefully before interpreting the peculiarities of the results of the national surveys or the critical discussion of its theoretical backgrounds. Duclos reached the following conclusion: 'We still absolutely don't know how people around the world really evaluate and envisage the environment. But we know a lot about what they probably don't think, or about what it would be interesting to know if we had enough means to dig into the 'cultural thickness'. We experience through such attempts the main paradox of many international comparisons (which is perhaps the paradox of most modelizations of hypercomplexity): that is, the 'realism' of artifacts depends on the sophistication of the means apt at capturing the 'cultural semantic universes', which, in turn, acquire maximum intelligible complexity at national levels, and goes probably beyond at the international level' (Duclos, 1993:70).

In doing a secondary analysis of Gallup's data, using the method of multivariate factor analysis and the concept of 'subjective rationality' I demonstrated that Dunlap's interpretation did not cover satisfactorily all the observed facts, and these findings need to be considered from different points of view (Rukavishnikov, 1994). The main point is that environmental views and behaviour have been shaped and guided largely by the wider cultural context, which embraced the overall value system of a given society. To find out *why* people have different environmental attitudes, *why* people behave in different ways and *what* prompts their views, scholars have to explore the link between social psychology, culture, laws and the stage of economic development of a given society. Explanation of people's views must be based on a broad socio-political and value reasoning. According to this perspective, questions of causes of environmental attitudes as the most important issue, cannot be separated from a consideration of the changes of the entire value system as well as economic and political transformations that underlie the process of development of public opinion regarding environmental issues in the given society.

This approach looks more analytical than descriptive though it has its disadvantages and limitations as well. The last are related to the skill of operationalization, which involves the translation of abstract concepts from verbal to measurable formulations, because due to specific socio-political and economic situations and cultural peculiarities of nations, it is impossible to use absolutely identical questionnaires in different countries.

Here we come across the methodological problem of the comparability of the value patterns. Value patterns seem to be comparable but not identical in every country. So, we need to know how value patterns spread in different nations, otherwise we cannot reason the observed differences in opinion that are revealed in surveys. Let us not forget that public opinion is defined also as the most common way of thinking in a given society that depends on both cultural traditions and past and present social circumstances.

What we know now

At the very beginning of the course of methodology, students of social sciences learn that 'the power of social research lies more in the realm of questions than in the realm of answers'. Starting from this fairly extreme position I have explored the set of basic research questions that are in the focus of attention of scholars in the field of interest of environmental sociologists. Now let us turn to the answers that raise new questions. For the sake of space I will focus only on some of them that are related to the above mentioned basic research questions. Summarizing the findings of studies carried out in the 1960's-1970's in the U.S.A., Mark W. Lipsey, in a chapter published in 1977, stated that 'the highest levels of environmental concern are generally found in the upper middle-class, among people who have high education, income, and occupational status. This is less true of those in business fields, however. The young express more environmental concern than their elders, but do not generally engage in more environmentally-relevant behaviour' (Lipsey, 1977:378-379). At that time rural people were usually less environmentally inclined than urbanites (Lowe and Pinkey, 1982). These conclusions were stated more than twenty years ago. A large majority of studies conducted later only confirmed these conclusions about the relationship between environmental attitudes and socio-demographic variables. As an example let us take again the article of Glen Filson, who studied the environmental attitudes of Ontario farmers in comparison with the views of their Australian counterparts in the 1990's. This scholar reported that 'environmental orientation was positively correlated with education and inversively correlated with age, income and acreage' and that 'the younger, well-educated female farmers in particular, were increasingly concerned about the seriousness of rural environmental degradation', but 'capitalist farmers were least environmentally oriented (except most retired farmers who were even less environmentally oriented)' (Filson, 1993:171,180).

Thus the statistical relationships between relevant variables as well as the structures of an environmentally oriented and concerned public *have not drastically changed* during the last thirty years, at least not in North America. Certainly, step by step, scholars got more and more data about the relationship between variables, but has the mountain of data helped us to achieve a better understanding of the ecological culture of North-Americans?

One can say that the fundamental aim of the descriptive kind of inquiry is a new quantitative kind of information. Does this kind of information have a heuristic value? Of course it has. The heuristic value of any descriptive findings depends on the degree of knowledge about the matter of interest. And, certainly, one of the simplest ways of economizing effort in an inquiry is to review and build upon the work already done by others. The important qualification in this regard is the general truth that everything we know today, every answer we have discovered yesterday, will be likely to be overturned at some point in the future, so the replication of former studies in changed circumstances has some sense. But how can we evaluate the value of such studies from the point of view of further development of explanatory sociological theory in the 1990's? I highly appreciate the classical scheme of the sociological interpretation of facts in terms of socio-economic and demographic variables, but I feel the reproduction of the expected and predicted conclusions about the relationship between variables, even when they are made on the new empirical basis, today cannot take us much closer to a better understanding of the nature of mass environmental consciousness.

Cross-national and cross-cultural studies based on polls carried out in dozens of countries all over the world, depending on issue and wording, showed a widespread concern about environmental deterioration and pollution. I am hesitant to regard this phenomenon as a sign of a global awareness of impending danger of global or local ecological disasters. For instance, people even kept on living in the Chernobil zone. In the situation when hardly any respondent is willing to admit publicly that he is not deeply concerned about the environment, we must raise the question whether this concern is produced by a real awareness of the seriousness of the problems (linked with the change of the old value structures into postmodern ones where environmental values are playing one of the dominating roles), or whether it is only a case of socially desirable responses, partly impacted by global media communications. Analyzing data of surveys conducted in 1990 in almost all the countries of Europe and in the US and Canada, Peter Ester and his colleagues came to the following conclusion:

'The message from the media is heard but not internalized. Ecological concerns are, for the greater part of the population, not yet part of a firmly rooted value system. The dangerous side-effects of economic growth and consumption patterns are seen and society wants to prevent them and restore the environment, but not at the cost of economic welfare. There has been no fundamental change as yet in attitudes towards nature' (Ester et al., 1993:180).

Looking back to the 1960's, we see a great shift in public concern regarding nature all over the world during the last three decades. Since the 1960's, when environmental issues came in the focus of public attention, the level of environmental concern has reached top scores. However, human nature has not changed during these thirty years, and the following basic conclusions that were made then are still relevant:

'The high general level of environmental concern does not mean, however, that most people are willing to suffer considerable personal effort, expense, or inconvenience to improve the environmental condition. Many expect improvements to come through technological innovations, not through changes in their way of life. People generally are willing to undertake a small amount of household conservation behaviour, but are relatively reluctant to accept an increased tax burden or other economic sacrifices to achieve environmental benefits. The public plainly feels that reduction of environmental problems is one of the top-priority responsibilities of the government' (Lipsey, 1977:379). Have the recent cross-national studies of the Gallup Institute, MORI and the European Value Study Group, that have confirmed these statements as relevant to most states, really gifted us with a new historic knowledge? I would like the answer to be positive. But, much as I would like to define cross-national data as the evidence of global changes of environmental attitudes and as confirmation of the modernization theory predictions, due to observed similarities of public worldviews and perceptions regarding environmental issues, there are good reasons to exercise caution before reaching such a conclusion.

The very fact that the publicly declared environmental concern of the majority of residents is not carried into public action and actual behavioral changes, including personal readiness to sacrifices and the individual's environment protection behaviour, is reflected in empirical data. Dutch scholars interpreted it as follows: 'The increase of concern and the fact that this concern is not yet firmly rooted in a value system, explains why neither socio-demographic variables nor more basic value orientations in other than ecological domains explain why some individuals are concerned about the environment and others are not, and why some are prepared to make sacrifices and others are not' (Ester et al, 1993:180). In my view this explanation is incomplete. It stresses the role of the changes of deeply rooted values according to the dogmas of the modernization theory, but underestimates the role of objective social, economic and political factors that put pressure on the societal value pattern and people's behaviour. Such cases can be found in developing, transitional and welfare states. For instance, the 'recent surveys of Canadian farmers have shown that most farmers are much more concerned about their economic survival than they are about the environment; even though 20% of Ontario farmers disagreed with the view that people who pollute are just as criminal as those who steal, only 7% of Australian farmers did' (Filson, 1993:166, 171). Data of my own Russian surveys in 1991-1994 have showed that even though most Russians have put the environment on the list of the most urgent problems of their country in the period of transition, they are much more concerned about sky-rocketing price risings, increase of crime and other issues linked directly to their everyday life and survival in the circumstances of radical social transformations running in this country. At the same time ordinary Russians have no reliable comparable objective instrumental information about the environment, so their evaluations are based more on emotions and media publications than on facts and values. To my mind, the influence of the economic, political and media factors, as well as the cultural factor, have to be accounted inside an entire unity in each country. Making this assumption, I, of course, do not insist that value variables have no significant impact on the public perception of environmental problems, but their significance in explanatory frames should not be overestimated.

Unquestionably, we cannot underestimate the growing role of globalization and the geo-cultural factor in producing the differences of the public opinion results in future cross-cultural perspective. During the last decades of the 20th century, the political and economic map of the world has changed very rapidly. Very soon, in the year 2000, it will be a new world with an inheritance of old environmental problems. The ongoing cultural processes of modernization and individualization are processes of transformation of a traditional value-attitude system into a modern one, that includes the increase of the relative significance of environmental values as an important ingredient. These processes are running all over the world. They run on parallel lines with the processes of economic and political development of the given country. However, the speed of changes of economic and political conditions of human activity is usually much higher than the pace of transformation of the entire value system. Here could be found some reasons for currently observed similarities and peculiarities of public opinion of the countries with different levels of economic development and culture.

4. CONCLUSIONS

I am far from the idea to oppose the well-developed old-fashioned descriptive approach and other strategies of the research orientation one to another, keeping in mind the advantages and limitations of each approach. Researchers cannot get complete answers to questions about causes without detailed descriptive information about the state of opinions and attitudes. The different scientific paradigms and methodological approaches are used now in single case studies as well as in national-scale, cross-national and/or cross-cultural studies, and will be used in the future. Our research instruments need to be improved (Worcester, 1994). Systematic tracking of public and elite attitudes to environmental issues is essential in the future if policy makers are to have an accurate understanding of public attitudes to the global and national environment. A free analogy with the art of painting may be useful for a conclusion of this discussion. In the art of painting it is easy to see both structural and representative elements. A picture is normally a picture of 'something': it depicts or illustrates a 'subject' made up of things analogous to 'objects' in sense experience. At the same time there are present certain elements of pictorial design: what a picture represents is organized into structural patterns and conventions which are found only in pictures. The words 'content' and 'form' in the art of painting are often used to describe these complementary aspects of painting. 'Realism' connotes an emphasis on what the picture represents; stylization, whether primitive or sophisticated (up to so-called non-objective painting), connotes an emphasis on pictorial structure. And so we may say without much fear of effective contradiction, that the whole art of painting lies within a combination of pictorial 'form' or structure and pictorial 'content' or subject.

By posing the appropriate questions, sociologists make a first step towards a picture of 'something'. In our case 'object' is the state of public opinion towards environmental problems and 'subject' is a picture of public opinion created by sociological tools. Questions are the main tool of social researchers just like a brush is to a painter. Survey research is just a method for obtaining specific information from a relatively small number of individual sources through questioning. I have not discussed the problems of data gathering and sampling here, but stressed the idea that the bias of results might be programmed by incorrect operationalization. When we regard survey research from a comparative perspective, we see that the wording of the questions used in polling, including that with the definitions of actual environmental problems, plays an extremely important role compared with the other elements of the study. Unfortunately, severe mistakes made on the stage of questionnaire design inevitably have an impact on the false inferences, and devalues the entire study. Questions and answers constitute a 'content' of the sociological picture. By using incorrect wordings one can create only a surrealistic picture of 'something' that looks like a reality ('object') but in fact this picture will be very far from it (so-called 'artifact').

The fundamental problem here is how to distinguish a 'surrealistic picture' of the given society at a particular time-point from a 'realistic' one. There are no precise prescriptions for this. Comparing findings of the given survey research with the results of other studies might be considered a reasonable way for obtaining the truth, but, naturally, it can only produce arguments and not evidences in any dispute. As I have said above, it is much more difficult to recognize the 'surrealistic landscape of views' in comparative studies.

However, can we, generally speaking, say that in the case of questions with perfect wording one will inevitably meet with a similar realistic picture of views and evaluations? Certainly not, because the content of questions depends on theoretical assumptions and concepts as well as on the knowledge of the current public needs and moods that finalized in the questionnaire design. One poll can display a picture of an overwhelmingly popular concern regarding the deterioration of the environment, and another poll, carried out simultaneously, might demonstrate that public awareness regarding environmental problems is comparably less than the public's concern about the worsening of the state of the economy in general or about personal economic survival.

Visible contradiction between these pictures - inferences from data of polls - might be misleading. One must be careful in judging methodological defects. Sometimes the alternative inferences revealed two different pictures of the same object ('public environmental concerns and attitudes'), created by different artists by different means and for different tasks. The heuristic value and political and practical significance of these insights may vary to a significant extent, because the cost of the picture in the eyes of the public or buyers is not directly relevant to its content as well as to the skill of the artist.

REFERENCES

Duclos, D. (1993). 'Some Remarks on the Gallup Survey "The Health of the Planet"'. *Sotciologicheskie Issedovania* 11:67-70 (in Russian).

Dunlap, R.E., G.H. Gallup Jr. and A.M. Gallup. (1992). *The Health of the Planet Survey. A preliminary report on attitudes towards the environment and economic growth measured by surveys of citizens in 22 nations to date.* The George H. Gallup International Institute, June.

Dunlap, R.E. and A.G. Mertig. (1994). 'Global Environmental Concern: A Challenge to the Post-Materialism Thesis'. Paper presented at the International Sociological Association's XIIIth World Congress of Sociology. Bielefield, Germany, July.

Ester, P., L. Halman and B. Seuren. (1993). 'Environmental Concern and Offering Willingness in Europe and North America'. In: P. Ester, L. Halman and R. de Moor (eds.), *The Individualizing Society: Value Change in Europe and North America.* Tilburg, Tilburg University Press, ch.8.

Filson, G.C. (1993). 'Comparative Differences in Ontario Farmers' Environmental Attitudes'. *Journal of Agricultural and Environmental Ethics* 6(2):165-184.

Lipsey M.W. (1977). 'Attitudes Towards the Environment and Pollution'. In Oskamp, Stuart (ed.), *Attitudes and Opinions.* Englewood Cliffs, N.J., Prentice-Hall, Inc.., Ch.16.

Lowe, G.D. and T.K. Pinkey. (1982). 'Rural-urban Differences in Support for Environmental Protection'. *Rural Sociology* 47:114-128.

Rukavishnikov, V. (1994). 'Is the Public Opinion on Ecological Problems Rational?'. *Sotciologicheskie Issedovania* 1:50-58 (in Russian).

Worcester, R. (1994). 'Societal Values, Behavior and Attitudes in Relation to the Human Dimensions of Global Environmental Change: Use of an Environmental Activist Scale'. Paper presented at the special session 24 'Comparing Public Opinion on the Environment', XVI IPSA World Congress. Berlin, 20-24 August.

About the authors

Alan Black is an Associate Professor of Sociology at the University of New England, Australia. He is a member of the International Steering Committee for the 1995-96 World Values Survey.

Peter Dickens is Reader in Sociology at the University of Essex, United Kingdom. He has carried out a wide range of research on social environmental issues. His book *Society and Nature* was published by Harvester in 1992.

Riley E. Dunlap is Professor of Sociology and Rural Sociology at Washington State University, and Gallup Fellow in Environment at the George H. Gallup International Institute. He is also currently serving as President of the International Sociological Association's Research Committee on Environment and Society (RC 24). Examining the nature and sources of public concern for environmental quality has been a major focus of his research since 1970.

Peter Ester is Professor of Sociology at Tilburg University and director of IVA, Institute for Social Research. He has published several books and numerous articles on environmental issues. His most recent books in English are *Social and Political Attitudes in Dutch Society* (co-author Paul Dekker; Den Haag: VUGA, 1993) and *The Individualizing Society* (Co-editors Loek Halman and Ruud de Moor; Tilburg: Tilburg University, 1994). He is chairman of the Dutch Association of Social and Cultural Sciences. His main research interest is in the field of comparative studies of values and value change.

August Gijswijt is environmental sociologist at SISWO, University of Amsterdam. He is secretary of the Research Committee on Environment and Society (RC 24) of the International Sociological Association.

Loek Halman is a Research Fellow at the Work and Organization Research Centre (WORC) of Tilburg University. He is secretary to the Steering Committee of the European Values Study. His main interests are comparative research on values and value change. He is involved in the European and World Values Studies, and, in collaboration with Nijmegen University, involved in the Dutch survey research project on Social and Cultural Developments in the Netherlands (SOCON). His main publications include *The Individualizing Society* (co-editors Peter Ester and Ruud de Moor; Tilburg: Tilburg University Press, 1993) and *De Cultuur van de Verzorgingsstaat* (co-editor Peter Ester; Tilburg: Tilburg University Press, 1994).

Irene Khalyi is researcher at the group on social ecological research of the Institute of Sociology, Russian Academy of Sciences. She is the author of numerous articles on social-ecological aspects of environmental movement relationships, and ecological aspects of national-patriotic movements in Russia.

Andrej Kirn is Professor at the Faculty of Social Sciences, University of Ljubljana, Slovenia. His main fields of interest are sociological and philosophical aspects of science and technology. He studies as well theoretical issues related to social ecology.

Angela G. Mertig is a Post-Doctoral Research Associate in the Department of Rural Sociology at Washington State University, where she recently received a Ph.D. in sociology. A specialist in environmental sociology and quantitative research methods, Mertig is co-editor (with Riley Dunlap) of *American Environmentalism: The U.S. Environmental Movement, 1970-1990* and co-author of several articles on environmentalism.

Andreas Metzner studied Social Sciences, Biology and Philosophy at the University of Münster which he finished with a doctorate in Philosophy. He is Assistant Professor in Environmental Issues in the Social Sciences at the University of Cottbus in Germany.

Elim Papadakis is Professor and Head of Sociology, University of New England, Australia. His principal interests are in environmental politics, social and political theory and political sociology. His most recent book, *Politics and the Environment. The Australian Experience* was published by Allen and Unwin in 1993. He is currently writing a book on *Environmental Politics and Institutional Change*.

Ian Reeve is a senior Project Director with the Rural Development Centre at the University of New England, Australia. He has recently commenced candidature in a Ph.D. research project 'Society, Culture and Waste: Household Practices and Public Policy'.

Vladimir Rukavishnikov is Head of the Department of Social Dynamics of the Institute of Socio-political Research of the Russian Academy of Sciences. He is a member of the Russian Academy of Social Sciences and an active member of the New York Academy of Sciences. He is also Professor of Sociology at Moscow State University and Vice-Editor-in-Chief of the journal *Sotsiologicheskie Issledovania* ('Sociological Studies') - the main academic sociological journal in Russia. Current research activities include social and political dynamics in Russia, value changes, public opinion and methodological issues.

Wolfgang Schluchter studied Sociology and Political Sciences at the University of Heidelberg where he completed his doctorate in Philosophy. He is currently Professor at the University of Cottbus in Germany where he holds the chair for Environmental Issues in the Social Sciences.

Brigitte Seuren studied political science and history at Nijmegen University, the Netherlands. She is a graduate student at Tilburg University, preparing a thesis on the relationships between value orientations and environmental concern and behavior.

Pieter Tijmes is a social and political philosopher at the Department of the University of Twente in Enschede, the Netherlands. He participates in the Research Program on Cultural Philosophy. His main research focus is on scarcity and technics.

Oleg Yanitsky is currently the head of the group on social ecological research of the Institute of Sociology, Russian Academy of Sciences. He is the author and editor of several books, the most recent being *Cities of Europe: the Public's Role in Shaping the Urban Environment* (1991, co-editor) and *Russian Environmentalism: Leading Figures, Facts, Opinions* (1993).